Tropical Cyclones

Tropical Cyclones

Climatology and Impacts
in the South Pacific

James P. Terry

 Springer

James P. Terry
School of Geography
The University of the South Pacific
Suva, Fiji Islands
james.terry@usp.ac.fj

ISBN: 978-1-4419-2447-6   e-ISBN: 978-0-387-71543-8

Printed on acid-free paper.

9 8 7 6 5 4 3 2 1

springer.com

For Daisy, Dylan and Joshua

Preface and Acknowledgements

The South Pacific is an almost incomprehensibly vast ocean. Within it lie thousands of islands belonging to over 15 developing nations and territories. These islands display enormous physical diversity. They may be large or small, rugged volcanic mountains with high elevations, flat limestone platforms or tiny coral islets resting just above sea level on top of coral reefs. Many islands are remote and uninhabited while others are densely populated with bustling towns and expanding cities. In terms of climate, the South Pacific is one of the major ocean basins where tropical cyclones occur. Most tropical islands are affected periodically by the passage of these violent storms, which cause loss of life, disrupt society and (my main interest) often produce spectacular changes in island physical environments. It is perhaps surprising then that no book has previously been dedicated to describing either the climatology of tropical cyclones in the South Pacific or their physical impacts on the islands they encounter.

The aim of this book therefore is to link two central themes – tropical cyclones and the physical environments of islands in the South Pacific. The first half of the book describes the characteristics and behaviour of tropical cyclones in the region, and assesses the outlook for the future in the context of climate change. The second half then illustrates the importance of these storms for island environments, concentrating on geomorphological and hydrological responses. Regional examples and case studies are used to show how coral reefs, coastlines, hillslopes and rivers are all affected, and how sometimes tropical cyclones can even cause the destruction of existing islands or the formation of entirely new ones.

It is certainly the case that plenty has already been written about tropical cyclones and a great many research papers may be found in scientific journals. But the language of these is not easily accessible to all. And so it shouldn't be, because the content is intended for specialist audiences – climatologists, meteorologists, physicists, 'tempestologists' and those from associated disciplines. My own experience, from over two decades of teaching physical geography at university, is that to try to cajole students without the necessary science backgrounds to read and use such materials is often a difficult task. Although most students are keen to learn, they tend to be too shy of revealing an inability, either real or perceived, to grasp fully the tricky mathematical and physical concepts underpinning the thermodynamic behaviour of our atmosphere and the processes leading to the formation of tropical cyclones. I sympathise with their plight.

With this in mind, I have written this book to be illustrative yet concise, and informative but non-technical. I hope that this makes it attractive to a diverse readership, especially to those interested in climate and climatic extremes, tropical islands, tropical environments, physical geography, geomorphology and the South Pacific region in general.

I am very much indebted to many people for their help in researching, writing, illustrating and producing this book. Foremost, I wish to thank Professor Cliff Ollier for allowing me to benefit from his enormous wisdom on writing a readable manuscript, and Professor Patrick Nunn for offering much sensible advice while reviewing the penultimate draft. Marie Puddister and Daisy Terry accomplished a great deal of hard work preparing early versions of many of the diagrams, for which I am grateful. A large number of other individuals provided much-needed assistance in the field, gave me unlimited access to their original unpublished data and photographs, shared their personal experiences or simply joined in useful and stimulating discussion. A few of those I would especially like to mention are Michael Bonte, Austin Bowden-Kirby, Ami Chand, Pradeep Chand, Douglas Clark, Antoine DeBiran, Finiasi Faga, Sitaram Garimella, Robert Gouyet, Tetsushi Hidaka, Kei Kawai, Ray Kostaschuk, Ravind Kumar, Riteshni Lata, Simon McGree, Rajendra Prasad, Rishi Raj, Nick Rollings, Roshni Singh, Randy Thaman, George Vakatawa, Aliti Vunisea and Geoffroy Wotling. To the many other people, too numerous to name individually, who helped in some way, I extend my sincere appreciation.

For giving me the inspiration and motivation to write this book, I am grateful to all my students in physical geography, both past and present, at The University of the South Pacific. The work presented herein would also not have been possible without the generous financial support of the University Research Committee.

During the last decade of fieldwork in my adopted home in the South Pacific, I have been fortunate enough to visit many islands and stay in traditional villages with the local people. On remote and isolated islands especially, daily life can be a struggle for the people who live there. Tropical cyclones certainly don't help. One has to admire the way Pacific islanders endure the physical challenges that such climatic hazards present. I offer my heartfelt thanks for the willing help, guidance, hospitality and companionship offered by all the Pacific islanders I have been privileged to meet.

J. P. Terry
Suva, January 2007

Contents

Chapter 5   Meteorological Conditions

Chapter 6   Future Tropical Cyclone Activity

Part II   Impacts of Tropical Cyclones

Chapter 7   Coastal Geomorphology

Part I
Tropical Cyclones in the South Pacific

Chapter 1
Setting the Scene

1.1 Introduction

The islands of the tropical South Pacific are vulnerable to a variety of natural hazards. Some are associated with the vagaries of climate, such as tropical cyclones (elsewhere called hurricanes or typhoons), droughts and floods, whereas others are geological in origin, such as volcanic eruptions, earthquakes and tsunamis. The main difference between these two groups of hazards concerns the timescale on which they happen. If we decide arbitrarily to deal on the timescale of a human lifespan, then the effects of a major geological event will be experienced perhaps only once or twice in a lifetime, or possibly not at all. In contrast, tropical cyclones and associated hazards are felt more frequently, perhaps every few years.

The aim of this book is to explore the characteristics and behaviour of tropical cyclones in the South Pacific and examine the remarkable impacts that these violent storms have on the physical environments of the islands they encounter. For the island nations of the South Pacific, tropical cyclones are nature's most intense phenomena, and their attendant conditions of tremendous seas, fierce winds, storm surge and torrential precipitation can inflict much damage on both human and natural landscapes. The combination of their severe nature and regular incidence means that tropical cyclones and their impacts are a major concern across the South Pacific; in short they are storms to be reckoned with for the people who live in this region.

It is something of a tradition in the opening sections of texts on tropical cyclones to quote a passage from the writings or diary of an early European explorer or missionary in the tropics who describes a historically severe event. Such excerpts are a vivid way to present an account of a tropical cyclone in more dramatic prose than the rather matter-of-fact language of science. The following example, recounted in Kerr (1976, p. 1), is a colourful account by the Reverend J. Williams (1837, p. 331–334), of a storm he experienced at Rarotonga in the Cook Islands:

The next day was the Sabbath, and it was one of gloom and distress. The wind blew most furiously, and the rain descended in torrents. . . . Towards evening the storm increased; trees were rent, and houses began to fall. . . . (Monday morning). . . . the whole island trembled to its very centre as the infuriated billows burst upon its shore.

Narratives of storms that are personal favourites of mine are the pair of excerpts highlighted by Visher (1925, p. 134), recollecting the tropical cyclone that struck Samoa on 16 March 1889. The tale of this storm is particularly astonishing as it is credited with having prevented war between the United States and Germany at the time! According to the Editor of the Independent newspaper of New York (1915), six warships of these nations were occupying the harbour of Apia, the Samoan capital, and were on the point of opening fire on each other when the storm hit (excerpt [1] below). The outcome of the cyclone's effect on the hostilities was assigned great significance on the turn of world history by the writer Robert Louis Stevenson (1915), who was a long-term resident of Apia (excerpt [2] below):

Then the storm broke. There were thirteen unlucky vessels afloat in Apia Bay when the sun rose. When it set, there were none. Twelve were sunk or grounded. One, the British warship "Calliope", had steamed out of the harbour mouth against the storm. If the battle had been fought, the loss of shipping could not have been greater. . . . Of the eighty Germans on the "Eber", only four were saved. When the news of the happenings reached Europe and America, the horror of it banished all thoughts of war. [1]

Thus in what seemed the very article of war, and within a single day, the sword arm of each of the two angry powers was broken; their formidable ships reduced to junk; the disciplined hundreds to a hoard of castaways. The hurricane of March 16 . . . directly and at once brought about the congress and treaty of Berlin; indirectly . . . it founded the modern navy of the United States. Coming years and other historians will declare the influence of that navy. [2]

At this point it is helpful to give a generalised definition for tropical cyclones, to avoid the possibility of any confusion with other types of climatic phenomena. Meteorology describes a tropical cyclone as a particularly violent type of migratory, non-frontal revolving storm, with low central barometric pressure, that forms over tropical waters. Because cyclones have steep pressure gradients, they incite phenomenal winds that drive enormous storm waves, and produce heavy condensation and torrential precipitation (Garbell 1947). The origin of the term 'cyclone' is based on the Greek word κυκλος ('kyklos'), meaning circle or coil, indicating the characteristically inward-spiralling flow of air towards the centre of the storm. A distinguishing feature of cyclones is their small, nearly circular area of clear and calm weather in the middle of the system, known as the *eye*.

Although these types of revolving storms are confined to specific regions across the globe, a variety of names exist according to the ocean basin within which they form. In the western North Atlantic and the Caribbean they are known as *hurricanes*; in the western North Pacific and China Sea as *typhoons*; in the western South Pacific and Indian Ocean they are called *tropical cyclones*. Throughout this book the term 'tropical cyclone' (or more simply 'cyclone') will normally be employed, often abbreviated to 'TC' where 'Tropical Cyclone' forms part of the name of an individual storm, for example 'TC Nancy'. The term 'storm' will also be used interchangeably with 'tropical cyclone', because other types of storms are not discussed.

The magnitude of the effects produced by tropical cyclones on the physical environments of islands in the South Pacific is a function of two sets of variables:

1. the severity of storms,
2. the sensitivity of the islands affected.

Storm severity is a wide-ranging term, used here to encompass all those variables related to the climatological characteristics and behaviour of tropical cyclone events. This includes storm frequency, intensity, speed of movement, longevity, size, proximity to islands, and so on. Island sensitivity, on the other hand, concerns the nature of different island types in the South Pacific, and how features of their physical geography influence and condition island responses to the various geomorphic and hydrological processes that are triggered during storm events. Owing to the great diversity of island types in the tropical South Pacific, geomorphic responses to tropical cyclones vary enormously. For example, the interior of volcanic islands with steep slopes and rugged terrain are sensitive to disturbances associated with slope and fluvial geomorphology, such as mass movements and river channel erosion. In comparison, low-lying coral islands with little relief built on atoll-reef foundations are sensitive to changes in their coastal geomorphology, for example reef damage and beach erosion.

The above division of the factors influencing tropical cyclone impacts on the physical geography of islands into the two groups given, namely storm severity and island sensitivity, is a simplistic model. Yet it does have the advantage that it provides a convenient framework for this book. Thus:

Part I (Chapters 1–6) explores and describes the climatology of tropical cyclones in the South Pacific.

Part II (Chapters 7–10) illustrates the physical geographic processes and landform changes that are the direct or indirect result of cyclone effects on islands.

1.2 The Study Area

The study area for this book is the tropical South Pacific Ocean. This is an enormous expanse of water, stretching across over 20 million square kilometres. Within this area lie thousands of islands belonging to more than 15 developing island nations, states and territories. From west to east these include Solomon Islands, New Caledonia, Vanuatu, Nauru, Kiribati, Tuvalu, Fiji, Wallis and Futuna, Tonga, Tokelau, Samoa, American Samoa, Niue, Cook Islands and French Polynesia, as shown in Fig. 1.1.

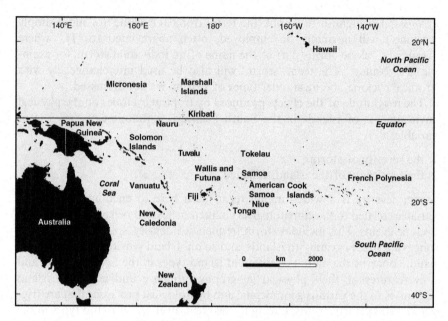

FIG. 1.1. The South Pacific Islands.

Conveniently, most of the study area falls under the responsibility of the Fiji Meteorological Service (FMS) for recording and archiving tropical cyclone activity. The FMS area of coverage extends from the Equator to Latitude 25°S and from Longitude 160°E to 120°W (Fig. 1.2). The FMS functions as a department under the Government of the Republic of the Fiji Islands (Fiji Meteorological Service 2006). The FMS headquarters is located in the compound of Nadi International Airport, in the town of Nadi on the western coast of Viti Levu island. Viti Levu is the main island in the Fiji group.

The FMS has two main output divisions, namely the Forecast Services Division and the Climate Services Division. The Forecast Services Division operates the Regional Specialized Meteorological Centre–Nadi Tropical Cyclone Centre (RSMC-Nadi TCC). The authority to operate as the RSMC for the tropical South Pacific region is granted by the World Meteorological Organization. The RSMC-Nadi is one of six Regional Specialized Meteorological Centres and an additional five Tropical Cyclone Warning Centres around the world. The areas of control and boundaries of these centres are shown in Fig. 1.2. The RSMC-Nadi provides weather forecasts, issues tropical cyclone warnings and other severe weather bulletins, and gives advisory meteorological information for the region. The RSMC-Nadi has the responsibility of naming and monitoring all tropical cyclones originating or moving into its region, and to issue warnings for the safety of all communities, including marine and aviation users.

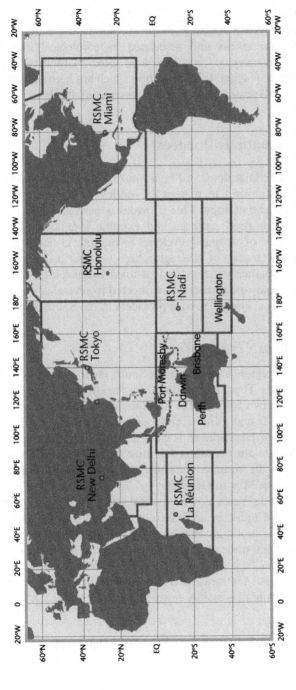

FIG. 1.2. Location of Regional Specialized Meteorological Centres (RSMC) and Tropical Cyclone Warning Centres (named), showing the geographical extent and boundaries of their regions of responsibility. Source: World Meteorological Organization WMO (2006a).

The second division of the FMS, the Climate Services Division, provides climate expertise and consultative services. It also serves as the national repository for climatological data and regional tropical cyclone activity records. The Climate Services Division of the FMS maintains the most comprehensive archive of historical tropical cyclone events and associated meteorological information (weather charts, satellite imagery, track data, wind speeds, barometric pressures, etc.) for the South Pacific region. It is this archive which has been utilised for the investigation of tropical cyclone characteristics presented in this book.

1.3 Regional Climatic Influences

Before describing the climatology and behaviour of tropical cyclones in the South Pacific, it is important to present an outline of some of the major controls on regional climatology. This provides a necessary yardstick for comparison between normal patterns and extreme events, and helps to explain many aspects of tropical cyclone formation in later chapters. Meteorological characteristics of the region are not described here, but for those interested in weather patterns and detailed meteorology of various island groups in the study area, readers are directed to several of the internet Web sites of the national climate service providers in the South Pacific.[1]

1.3.1 The Southeast Trade Winds

Most of the western region of the tropical South Pacific benefits from the Southeast Trade Winds. These are produced by the effect of Coriolis deflection on surface air drawn towards the low pressure region at the Equator, called the Equatorial Trough. The Coriolis deflection of winds, which is to the left in the Southern Hemisphere, is caused by the Earth's rotation. The Southeast Trade Winds are persistent for most of the year, although they tend to be weaker in the summer season (from November to April), and stronger in winter (from May to October).

As the Southeast Trade Winds blow across vast stretches of open ocean, they collect large amounts of moisture derived from evaporation at the sea surface. Vertical mixing of this moist air may give rise to some condensation and clouds. In general though, trade wind weather over the atolls and the other low islands of southwest Pacific is clear and fresh. This is because the lack of significant relief on low islands means that no clouds form by the process of orographic lifting, so rainfall is mainly convectional and frontal. This situation

---

[1] For Fiji: http://www.met.gov.fj; for French Polynesia: http://www.meteo.pf; for New Caledonia http://www.meteo.nc; for New Zealand: http://www.niwascience. co.nz; for Samoa: http://www.meteorology.gov.ws; for Tonga: http://www.mca.gov. to/met; for Vanuatu: http://www.meteo.vu.

is in marked contrast to high volcanic islands, where the orographic effect is the most important rainfall-generating mechanism on the windward southeast sides of islands facing into the Trade Winds. This can result in big geographical variation in annual rainfall totals across many high islands, and local people often refer to the 'wet' and 'dry' sides of their islands.

The interaction of volcanic relief and the Southeast Trade Winds on rainfall distribution is illustrated in Table 1.1 for Viti Levu island in Fiji. Suva city on the southeast coast of Viti Levu lies on the wet side of the island, and receives almost 3,000 mm of rain annually during 240 rain days. Lautoka city on the northwest coast enjoys a drier location in the lee of the volcanic highlands in the interior of the island. Lautoka therefore receives 1,903 mm of rain a year in less than half the number of rain days experienced by Suva. Relative humidity is correspondingly higher in Suva than in Lautoka, and as might be expected, this has an inverse effect on daily sunshine hours for the two cities. Isohyets extrapolated across entire Viti Levu are seen in Fig. 1.3, showing that the wettest part of the island is the central volcanic highlands.

TABLE 1.1. Long-term climatic averages for two coastal sites in Fiji. Suva and Lautoka cities are located on the southeast windward and northwest leeward coasts of Viti Levu Island, respectively.

Suva City

Station location: Laucala Bay; Latitude 18°09S Longitude 178°27E; Elevation: 6 m

Jan	Feb	Mar	Apr	May	Jun	Jul	Aug	Sep	Oct	Nov	Dec	Year
Rainfall in millimetres (1942–2002)												
343	293	373	366	245	168	142	148	191	206	254	269	2998
Number of rain days (1942–2002)												
23	22	24	22	20	18	18	17	17	19	19	21	240
Relative humidity at 9 a.m. in percent (1942–2002)												
81.3	82.4	83.5	83.2	81.6	82.3	80.4	79.7	79.3	78.3	78.8	79.3	80.8
Sunshine hours per day (1926–2002)												
6.1	6.0	5.5	5.1	4.7	4.5	4.4	4.7	4.4	5.1	5.6	6.2	5.7

Lautoka City

Station location: Lautoka Sugar Mill; Latitude 17°37S Longitude 177°27E; Elevation: 19 m

Jan	Feb	Mar	Apr	May	Jun	Jul	Aug	Sep	Oct	Nov	Dec	Year
Rainfall in millimeters (1910–2002)												
302	326	338	181	99	65	52	68	73	89	124	186	1903
Number of rain days (1910–2002)												
16	16	17	12	7	5	4	5	6	7	9	12	116
Relative humidity at 9 a.m. in percent (1957–2002)												
74.3	76.5	77.0	75.5	74.1	74.2	71.7	69.2	68.7	68.6	69.0	70.3	72.4
Sunshine hours per day (1957–2002)												
6.7	6.7	6.8	6.4	6.9	6.8	7.1	7.5	7.0	7.5	7.4	7.4	7.0

They are separated by volcanic mountains that rise in the centre of the island to over 1,300 m above the sea level. Data Source: Fiji Meteorological Service.

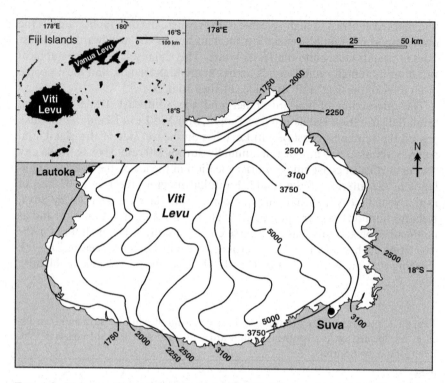

FIG. 1.3. Isohyets of annual precipitation on Viti Levu, the main island in Fiji, illustrating the difference in rainfall between the wet windward and drier leeward sides of the Island. The pattern is produced by the orographic effect of the high interior volcanic relief on the dominant Southeast Trade Winds. Source: Fiji Meteorological Service.

1.3.2 The South Pacific Convergence Zone

The second main regional climatic influence is the South Pacific Convergence Zone (SPCZ). This is a wide band of low pressure with an approximate northwest to southeast orientation over the southwest Pacific, extending diagonally from near Solomon Islands, across to Samoa, the Cook Islands and beyond (Salinger et al. 1995). The SPCZ marks the boundary between the Southeast Trade Winds and the Divergent Easterly Winds farther to the northeast, produced by a high pressure system that sits over the eastern part of the southwest Pacific on a semi-permanent basis. Since the SPCZ is a low-pressure trough, it is associated with cloud and rain.

An important feature of the South Pacific Convergence Zone is its seasonal migration (Hay et al. 1993, Vincent 1994). It generally lies equator-wards, i.e. to the north of its average position, in mid-winter (July), and moves to occupy a more southerly position by mid-summer (January) (Fig. 1.4). During the summer, the SPCZ tends to be better defined and have

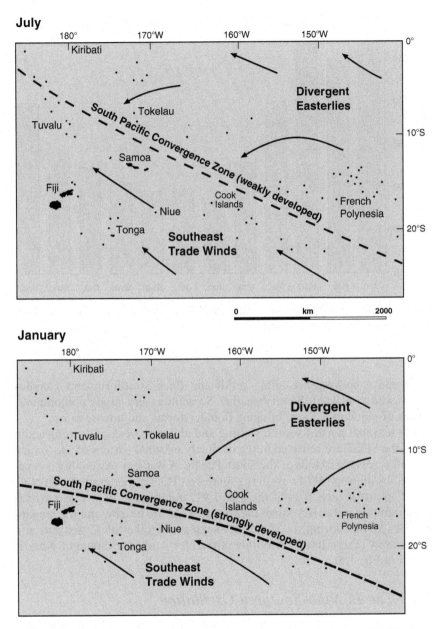

FIG. 1.4. Seasonal migration of the South Pacific Convergence Zone (SPCZ) from mid-summer (January) to mid-winter (July). Adapted from Nunn (1994).

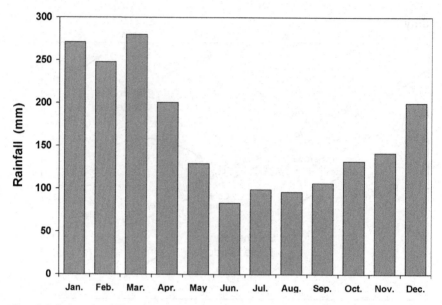

FIG. 1.5. Long-term (1905–1999) monthly rainfall for Alofi, the capital of Niue. Data Source: Niue Meteorological Office.

more active convergence, often producing thick stratiform and cumulus clouds, and associated showery weather. Sometimes very large cumulonimbus towers of cloud may form, bringing thunderstorms and intense rain.

The seasonal north-to-south shifting and alternate weak and strong activity of the SPCZ are reflected in the distinctly seasonal pattern of the annual rainfall across the islands of the South Pacific. A wet–dry seasonality is experienced by all tropical islands in the southwest Pacific. For example, on Niue island, approximately 67% of the total 1,992 mm of rainfall in a year arrives in the summer wet season when the strong SPCZ lies nearby, and the remaining 33% arrives during the winter dry season when the SPCZ weakens and moves away (Terry 2004a). Monthly rainfall for Alofi, the capital of Niue, is shown in Fig. 1.5.

1.3.3 The El Niño-Southern Oscillation

A third major control on climate of the South Pacific is the El Niño-Southern Oscillation (ENSO). At the inter-annual timescale, the ENSO phenomenon is our planet's most powerful climatic influence (Hilton 1998). Under normal conditions, low pressure at the Equatorial Trough and high pressure in the eastern Pacific establish a pressure gradient that keeps the Southeast Trade Winds blowing strongly. The combination of the Southeast Trades and the South Equatorial Ocean Current flowing east to west below the Equator allows the build-up of a very large body of warm water in the western equatorial Pacific, centred north of Australia and New Guinea.

At intervals of about 5–7 years, for reasons climatologists and oceanographers do not yet fully understand, there is a major disturbance to the coupled Pacific ocean–atmosphere system. This is called a positive ENSO anomaly or El Niño event[2] and can last for more than a year. An El Niño

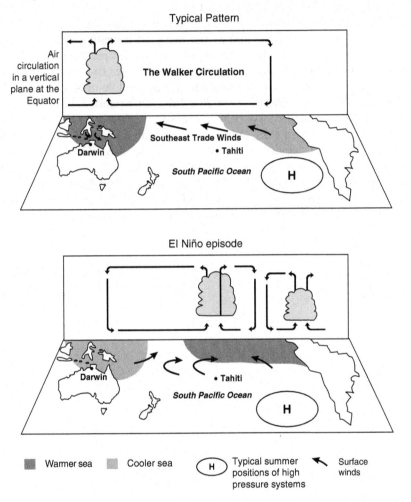

FIG. 1.6. Changes in atmospheric circulation, sea temperatures and surface winds from normal to El Niño conditions across the South Pacific Ocean. Source: Australian Bureau of Meteorology.

---

[2] El Niño refers properly to the oceanic component of the El Niño-Southern Oscillation system, the Southern Oscillation refers to the atmospheric component and ENSO refers to the coupled ocean–atmosphere system. In practice, El Niño is sometimes used to refer to the entire system (NOAA 2006a).

event begins with a Southern Oscillation, which is a shift in the normal large-scale atmospheric pressure pattern across the Pacific. The low-pressure zone centred north of Australia is replaced by high pressure, and the equatorial-zone pressure falls below normal. This causes the Southeast Trade Winds to lose strength. Without these winds to hold it back, the large pool of warm ocean water around New Guinea surges slowly eastwards across the equatorial Pacific (Congbin Fu *et al.* 1986) (Fig. 1.6).

At the onset of El Niño events, convective storms affecting the South Pacific islands may be generated as the eastward-migrating pool of warm ocean water passes across the north. As sea-surface temperatures rise off the western coast of the Americas, a tongue of warm water stretches back along the Equator. Rainfall becomes abundant in this new low-pressure region. In contrast, as El Niño conditions develop fully, an equatorward shift in the SPCZ leads to prolonged dry spells in the western South Pacific, and many island groups suffer rainfall failure and drought to varying degrees. The strength of ENSO activity is expressed as the Southern Oscillation Index (SOI), which is a measure of monthly atmospheric pressure differences between Tahiti and Darwin (Ropelewski and Jones 1987, Allan *et al.* 1991). Many climatologists think that the El Niño events that occurred in 1982–1983 and 1997–1998 were the strongest of the last century.

Chapter 2
Tropical Cyclogenesis

2.1 Principles and Controls

The initiation, development and subsequent maturation of tropical cyclones is known as *tropical cyclogenesis*. The detailed nature of tropical cyclogenesis continues to be under investigation by climatologists, and it is apparent that there is still much we need to learn. This section therefore concentrates on the primary reasons why and main processes how tropical cyclones form in the South Pacific region. In order to be able to understand these fundamentals, it is necessary to have a good grasp of how rising air masses undergo adiabatic cooling, which leads to condensation of water vapour and the simultaneous release of latent heat energy. The adiabatic process is not described here because existing texts on meteorology and physical geography give clear and comprehensive explanations (Barry and Chorley 1998, Strahler and Strahler 2006).

To start off, a helpful approach is to consider the tropical cyclone system as an 'atmospheric heat engine', fuelled by evaporation of warm water at the ocean surface. The underlying principle is that tropical cyclones originate from a state of thermodynamic disequilibrium that exists between the ocean and the atmosphere. The energy that tropical cyclones release is therefore the result of an attempt by the coupled ocean–atmosphere system to attain thermodynamic equilibrium, by the transfer of heat from the warm sea to the relatively cool atmosphere above, through convection and condensation. The incredible violence of tropical cyclones can thus be attributed to the enormous quantities of heat energy released by the condensation of water vapour contained in convectively unstable, moist and rising tropical air masses (Garbell 1947).

An important factor on which storm formation is dependent is a high moisture content in the mid-troposphere. This is necessary because elevated relative humidity levels in the surrounding atmosphere allow more rapid saturation and condensation of moist and rising air masses than would otherwise occur. If the mid-tropospheric air were instead relatively dry, then the atmosphere would be able to 'soak up' much of the additional moisture being supplied as water vapour, without condensation occurring.

A favourable atmospheric environment is also crucial, meaning that the atmosphere should be characterised by little disturbance, with minimal horizontal and vertical *wind shear* patterns (Revell and Goulter 1986a). Vertical wind shear is the gradient of wind velocity with height. Little change in wind velocity with height gives a low wind shear pattern, and conversely large changes in wind velocity up through the atmosphere cause strong wind shearing. The amount of vertical shear depends on the temperature structure of the air (Barry and Chorley 1998). If there are conditions of strong wind shearing in the atmosphere, this upsets the vertical structure of the tropical cyclone and prevents the maintenance of air outflow at the top of the system. Indeed, many tropical cyclones fail to develop properly or mature to their full potential intensity as a result of shearing.

2.2 Storm Formation and Development

2.2.1 Early Stages

Tropical cyclones develop initially within regions of low pressure and from pre-existing tropical disturbances. In the eastern region of the South Pacific Ocean, the Equatorial Trough of low pressure rarely, if ever, penetrates south of the Equator. This is one constraint preventing the formation of tropical cyclones in this part of the South Pacific. By comparison, in the western South Pacific, the South Pacific Convergence Zone (SPCZ) described in the previous chapter is a persistent and often well-developed regional zone of low pressure. The SPCZ is therefore often responsible for the initiation of tropical depressions (Revell 1981), and incipient tropical cyclones are often observed embedded within it. Figure 2.1 shows an example of a strongly defined SPCZ on 7 March 2000, extending from the Solomon Islands across Tuvalu, Samoa and down to Niue. At the southeastern end of the SPCZ is an embedded low-pressure system and associated cloud mass, which soon afterwards became organised into Tropical Cyclone Mona.

In the formative stage of a cyclone, an unusually active but poorly organised area of disturbance and convection appears on satellite images. The circulation centre is normally ill-defined and the strongest surface winds tend to occur in disorganised squalls (Australian Bureau of Meteorology 2006). A large area of ocean with surface temperatures around 27°C or above is necessary to thermally 'kick start' a better-defined cell of strong convection at the centre of the low-pressure area. This cannot normally occur in the southeast Pacific because along the western coast of the South American continent, and up to a thousand kilometres offshore, the upwelling waters of the cold Humboldt Current originating from the Antarctic Ocean keep the sea-surface temperatures below this threshold, effectively preventing tropical cyclogenesis. Farther west in the South Pacific, however, the water temperatures increase as the South Equatorial Current flows westwards south of the Equator, gradually warming up by insolation.

FIG. 2.1. Visible image from a geostationary meteorological satellite at 11:30 a.m. on 7 March 2000 (UTC[1]), showing cloud formation organised along the South Pacific Convergence Zone. The cloud mass just west of Niue is the initial phase of what became Tropical Cyclone Mona on 8 March and affected the Kingdom of Tonga. Base image courtesy of the Japan Meteorological Agency.

Once a tropical low-pressure system has an active convective cell established at its centre, there is a difference in atmospheric pressure between the middle of the depression and the surrounding area of relatively higher pressure. This difference causes a pressure gradient to be set up, and air is drawn inwards. If the minimum surface pressure drops rapidly, convergence intensifies. Strong winds are generated, with the warm ocean surface providing the moving air with heat and moisture through evaporation (Emanuel 1987). The maximum winds are now concentrated in a tight band close to the low-pressure centre, instead of the earlier disorganised pattern.

As the converging air is drawn inwards, the effect on wind direction of the *Coriolis Effect*, associated with the rotation of the Earth, becomes important. The Coriolis Effect causes the winds to be deflected from a straight line path as they are drawn inwards by the central low pressure. According to Ferrel's Law, the direction of the wind deflection is to the left in the Southern Hemisphere. The Coriolis Effect is therefore responsible for the spiralling motion around the central vortex of the depression (Fig. 2.2) and is the main reason that tropical cyclones develop their classical rotational nature.

---

[1] UTC means Universal Coordinated Time and is equivalent to Greenwich Mean Time or GMT.

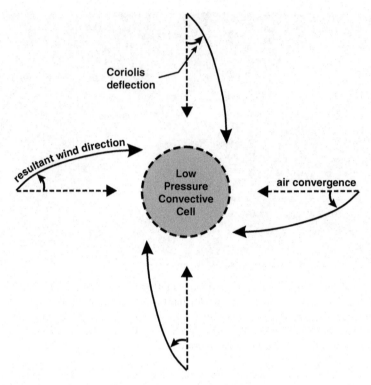

FIG. 2.2. Coriolis Effect influencing the direction of air movement into the centre of a nascent tropical cyclone, leading eventually to the organisation of the winds into the classical spiral pattern.

The strength of the Coriolis Effect, called the Coriolis Force, is negligible at the Equator and increases towards the poles. In consequence, approximately 74% of tropical cyclones in the South Pacific develop in latitudes 10–20°S (Table 2.1). The weak Coriolis Force near the Equator means that few tropical cyclones can develop in the zone from the Equator to 5°S, in spite of the presence of warm ocean water providing suitable conditions for convection. Beyond 20°S the presence of cooler ocean prevents sea-surface temperatures attaining the 27°C threshold for active convection over a wide enough area.

TABLE 2.1. Latitudes of tropical cyclone formation in the South Pacific, east of 160°E (1970–2004 data).

Latitudinal zone of origin	Percentage of tropical cyclones
0–5°S	1
5–10°S	19
10–15°S	48
15–20°S	26
20–25°S	6

2.2.2 Mature Stage

As the converging air spirals inwards, it becomes organised into bands of cloud that rotate slowly clockwise around the storm centre. The spiral bands of cloud are known as *feeder bands* because they feed heat and moisture into the central low pressure. This is essential for the continuing development of the cyclone, by providing energy into the storm system. At a certain distance from the centre of the low pressure, typically 20–40 km, the inflowing air suddenly turns upwards in a ring of intense uplift surrounding the eye (Fig. 2.3). This is called the *eyewall* and is the ring of strongest winds and heaviest precipitation. The upward transport of heat and moisture in the eyewall produces the vertical growth of immense thunderstorm clouds. The thunderstorms release large amounts of heat at mid and upper levels of the troposphere through the condensation of water vapour, allowing clouds to grow to the very top of the troposphere.

Once the tropopause is reached, the presence of the thermal inversion at this level in the atmosphere causes the air and clouds to spread out horizontally. This spreading pattern is known as the upper-level *divergence* or anticyclone. The divergence at the summit of the storm, at a height of between 9,000 and 15,000 m in the upper troposphere, appears to be another key process on which continuing cyclone development depends. Divergence aloft is essential to enable the high-level outflow of all the air that is being drawn in by convergence at the base of the storm. In effect, the upper-level divergence 'ventilates' the system. Most heat energy is exported by the anticyclone

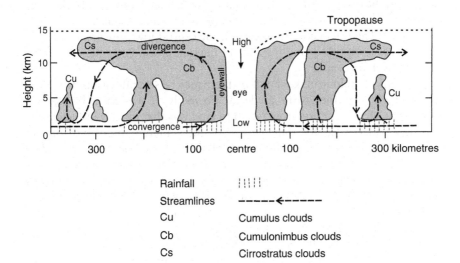

FIG. 2.3. Cross-section through a tropical cyclone to illustrate size, cloud structure, eye and streamlines of air movement. Modified from Barry and Chorley (1982).

circulation at an altitude of about 12 km. At this altitude the atmospheric pressure is approximately 200 mb, so climatologists refer to divergence occurring in streamlines at the 200 mb layer.

The upper divergence pattern also establishes an important positive-feedback mechanism. By intensifying the central low pressure at sea level and enhancing the low-level convergence pattern, this continues the cycle to fuel the system. During this stage a tropical cyclone is considered to be mature and has acquired a quasi-steady state, with only random fluctuations in central pressure and maximum wind speed.

2.3 Storm Decay

2.3.1 Major Influences

All tropical cyclones eventually weaken and decay. This is because the maintenance of storm intensity depends on the following essential ingredients:

1. Supply of warmth at the sea surface,
2. Supply of moisture at the sea surface,
3. Release of latent heat through condensation in the mid-troposphere,
4. Removal of air aloft by upper-level divergence.

If one of these four key ingredients is removed, the storm inevitably dies down. Tropical cyclones may decay over either land or ocean, and in either tropical or extra-tropical latitudes. The decay phase is observed in satellite images by the disruption of organised convection near the centre of the storm and the disappearance of the major curved bands of cloud farther out, although low-level circulation may remain well defined by narrow bands of low clouds. The warm core is destroyed, the central low pressure at sea level rises as the storm fills in and the zone of maximum wind expands away from near the cyclone centre.

If a tropical cyclone makes landfall, there may be several reasons why there is a tendency for the storm to decay. Formerly, it was thought that the increase in surface friction from ocean to land was the primary influence, but it is now widely accepted that the loss of the source of moisture from the warm ocean surface is more important (McGregor and Nieuwolt 1998). This is because any decrease in evaporation over the land surface is coupled with a dramatic reduction in the amount of latent heat released through moisture condensation at higher levels in the atmosphere (Anthes 1982). Many tropical cyclones forming in the Coral Sea west of 160°E move over the Australian continent and then die out for this reason.

With regard to the effects of cyclones encountering islands, the influence of low limestone islands and coral islands on atolls in the South Pacific appears to be of little significance in triggering cyclone decay. Intuitively this seems reasonable considering the small size, minimal elevation and subdued

TABLE 2.2. Examples of tropical cyclones that survived a traverse across a high island.

Island	Country or territory	Maximum elevation (m)	Tropical cyclone	Date
San Cristobal	Solomon Islands	1040	Emily	Late March 1972
Santa Isabel	Solomon Islands	1219	Ida	Late May 1972
Grande Terre	New Caledonia	1628	Alison	7 March 1975
Espiritu Santo	Vanuatu	1879	Marion	12 January 1977
Vanua Levu	Fiji	1032	Tia	17 March 1980
Viti Levu	Fiji	1324	Wally	5 April 1980
Savai'i	Samoa	1858	Val	7 December 1991
Pentecost	Vanuatu	947	Esau	26 February 1992

topography of these types of islands. A more interesting observation, however, is that plenty of examples exist where tropical cyclones have survived direct traverses of large high islands with elevations above 1,000 m. Table 2.2 provides examples of tropical cyclones crossing eight high islands in Fiji, New Caledonia, Samoa, Solomon Islands and Vanuatu, without significant weakening. The examples given seem to contradict the evidence that high islands can seriously disrupt air flow and convergence, leading to the initiation of storm decay. This observation has been made from the passage of hurricanes over mountainous islands in the Caribbean Sea (Pielke and Pielke 1997). At least two interpretations are possible for this. The topography of South Pacific islands may be below some threshold in elevation necessary to trigger cyclone decay. Alternatively, the temperature of the ocean surrounding the islands was sufficiently warm to sustain the moisture and energy requirements of these particular cyclones during their island traverse.

Nonetheless, land influences do include a cooling of the lower layer of the storm and increasing surface roughness compared to the relatively smooth ocean. The friction over land acts both to decrease the sustained winds and also to increase the gustiness felt at the surface. The sustained winds are reduced in strength because of the dampening effect of roughness over vegetation and buildings, but the gusts are stronger because of greater turbulence in the lower part of the storm, which has the effect of bringing faster winds down to the ground surface in short bursts.

The absence of any very large insular land masses ($>30,000$ km^2) in the tropical South Pacific, plus the extremely low density of islands compared to the vast expanse of the surrounding ocean, means that the majority of tropical cyclones die out over open water. For those cyclones which survive long enough to progress into extra-tropical latitudes, their decay is simple to explain, being a consequence of the deprivation of warmth required to fuel the heat engine of the storm system. Heat deprivation results from cooling of the base of the tropical cyclone. This occurs both by contact with cold ocean waters and by the invasion of cold air masses from higher latitudes, drawn in by convergence.

Yet, many cyclones weaken while remaining over warm ocean waters in tropical latitudes, and for these storms another reason must be provided to explain their demise. The most common reason for cyclone decay over warm tropical waters is that the storm track enters a region experiencing strong vertical shearing. A high shear environment disrupts the vertical structure of the tropical cyclone, breaking the link between the surface wind convergence and the upper-tropospheric wind divergence (Fig. 2.4). In extreme cases, strong vertical shear can tear a tropical cyclone apart (Fig. 2.5).

The subtropical *jet stream* in the South Pacific is often the cause of upper-level shearing. The jet stream is a narrow belt at high altitude in the upper troposphere in which winds reach great speeds, bounded by relatively stagnant air. The jet stream normally shows a wave-shaped pattern from above, but the overall wind direction is generally westerly (from the west). The South Pacific subtropical jet stream tends to occupy a position at a lower latitude (closer to the Equator) than the equivalent jet stream in the North Pacific. This helps to explain why fewer South Pacific tropical cyclones travel farther south than 30°S, compared to typhoons in the North Pacific, many of which maintain their structure and intensity far into northern subtropical and even temperate latitudes well beyond 30°N, regularly affecting the Korean Peninsula and the islands of Japan.

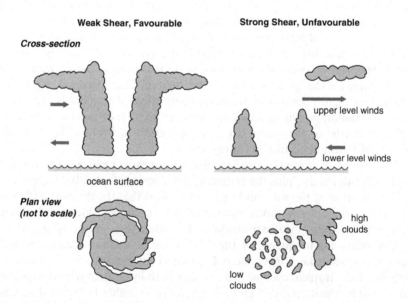

FIG. 2.4. Disruptive effect of atmospheric wind shear on tropical cyclone structure. Adapted from Pielke and Pielke (1997).

FIG. 2.5. Decay of Tropical Cyclone Nancy on 16 February 2005 while located in the southern Cook Islands. The satellite image shows circulation around the weakening vortex, indicated by the arrow. The dense upper-level cloud mass is shearing off to the east of the system. Base image courtesy of NASA.

2.3.2 Case Study – Decay by Vertical Shearing of Tropical Cyclone June in May 1997

A good example of cyclone decay by vertical shearing is that of Tropical Cyclone June in May 1997. The initial tropical depression developed close to the main Fiji Islands, being officially named TC June at a location of 14°S 174°E early on 3 May. According to historical data since 1840, TC June has the distinction of being only the fourth cyclone to threaten Fiji outside the normal cyclone season between the months of November and April (Fiji Meteorological Service 1997a). The life and behaviour of this system are unusual in several ways, reflecting TC June's development at an unusually late time of the season, when sea temperatures and climatic conditions are not as conducive to sustaining a tropical cyclone as earlier in the wet season (Terry and Raj 1999). The Fiji Meteorological Service (1997a, p.1) reported that *"TC June was a midget cyclone that never really formed a visible eye due to an insufficiently favourable environment."*

During most of its short three-day life, TC June remained relatively feeble, strengthening only to gale intensity (winds at 34–47 knots or 63–87 km h^{-1}), except for approximately 24 h from the early morning of 4 May when the system intensified into the storm force category (winds 48–63 knots or 88–117 km h^{-1}). TC June displayed erratic track behaviour, making sudden changes in both direction and speed that the weather forecasters at the RSMC-Nadi in Fiji found difficult to predict. In particular, after moving slowly but steadily on a southeast course towards the main Fiji archipelago during 4 May, TC June then decelerated and remained almost stationary to the northwest of the Yasawa islands group the next day, before taking an unexpectedly sharp turn southwards (Fig. 2.6). TC June began to lose structure, weaken and die out while still in Fiji waters, moving slowly westwards away from the coast of Viti Levu island. The main characteristics of TC June are summarised in Table 2.3.

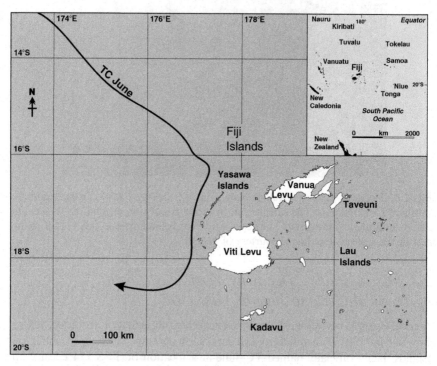

FIG. 2.6. Track of Tropical Cyclone June through Fiji waters from 3 to 5 May 1997.

TABLE 2.3. Summary of the characteristics and behaviour of Tropical Cyclone June, which affected the Fiji Islands from 3 to 5 May 1997.

Parameter	Details
Origin	14°S 174°E
Timing	Unusual – outside the normal cyclone season
Development	Indistinct eye formation
Size	Small – 'midget' classification
Speed	Travelled slowly, average 13 km h⁻¹
Movement	Erratic acceleration and deceleration
Direction	Showed unpredictable changes in direction
Intensity	Weak – remained mostly at gale strength
Lifespan	Short – 3 days
Track	Short – decayed while in tropical waters
Decay	Caused by vertical shearing

TC June experienced strong vertical shearing on 4–5 May 1997, which coincided with the onset of its decay. Upper layers of the storm, and associated cloud-bearing rain, were pushed east over the Lau group of the Fiji Islands, whereas the centre of the lower-level vortex remained to the northwest of the Viti Levu mainland (Fig. 2.7). Studying the position of the cyclone track in isolation from other climatic characteristics would therefore give a misleading impression of areas most affected by TC June, because the mass of the cyclone's rain-bearing cloud shifted to the east of the track. This gave an unexpected distribution of high-intensity rainfalls across eastern Fiji during the later period in the life of the storm, and residents in many places on Taveuni and Vanua Levu islands were taken by surprise when localised flooding occurred.

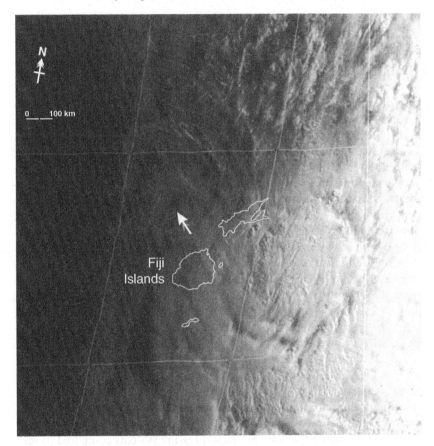

FIG. 2.7. Vertical shearing of Tropical Cyclone June on 5 May 1997 (6:30 a.m. Fiji Standard Time[2]). The satellite image shows clearly the exposed low-level vortex circulation located to the north of Viti Levu island, indicated by the arrow, and the upper-level cloud mass shearing off rapidly to the east. Base image courtesy of NOAA and the Fiji Meteorological Service.

---

[2] Fiji Standard Time is 12 h ahead of Universal Coordinated Time (UTC + 12 h or GMT + 12 h).

Chapter 3
Tropical Cyclone Structure

3.1 Shape and Size

Atmospheric pressure at sea level at the centre of a tropical cyclone is frequently as low as 965 mb. Away from the centre the pressure increases to about 1,020 mb at a storm's outer edge. This spatial variation in atmospheric pressure at sea level means that one way of examining the shape and size of a tropical cyclone (as with other types of weather systems) is to observe the arrangement of isobars on a synoptic weather chart. The isobar pattern displayed when a storm is slow moving or stationary is often a neat, nearly circular arrangement of concentric rings. This is because there are none of the fronts commonly seen in the structure of mid-latitude depressions (McGregor and Nieuwolt 1998). More rapidly moving tropical cyclones commonly show elliptical or pear-shaped patterns in their isobars (Fig. 3.1). Any elongation in shape is normally oriented in the direction of the storm track (Visher 1925) with the ratio of longest to shortest diameter about 3:2. In elliptical and pear-shaped patterns the isobars are not concentric. There tends to be some bunching of isobars in the leading portion of the storm, relative to storm movement along its track, and some spreading of isobars in the wake. Figure 3.1 illustrates this effect in the isobar patterns of Tropical Cyclones Olaf and Nancy near Samoa and the southern Cook Islands in February 2005.

The diameter of tropical cyclones averages 500–700 km across. Their area, therefore, covers approximately 300,000 km². Although this is a large area, tropical cyclones are relatively small low-pressure systems compared to mid-latitude depressions, which are often twice as big. Another way of measuring tropical cyclones is in degrees, radially outward from the centre to the outermost closed isobar. The distance in degrees determines the storm's size according to the widely used classification system in Table 3.1. Most South Pacific tropical cyclones fall into the 'medium to average' category, with a radius between 3° and 6°.

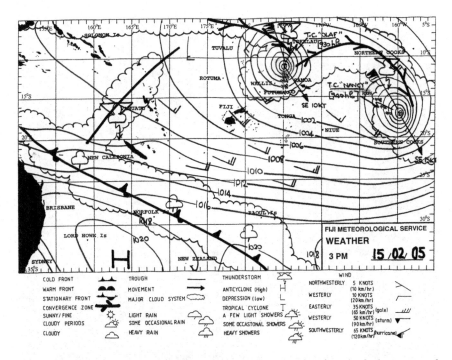

FIG. 3.1. Synoptic weather chart of 15 February 2005 showing the isobars around Tropical Cyclones Olaf and Nancy, which are both moving along southeasterly tracks. Note that the shape of the two storms is elongated, with alignment in the direction of movement. This is caused by bunching up of isobars in advance of the storms and spreading out of isobars behind. The shape of TC Nancy is more ellipti-cal and that of TC Olaf more circular. This is owing to the faster speed of TC Nancy, travelling southeast at 15 knots (28 km h^{-1}), compared to TC Olaf's slower progress at 10 knots (19 km h^{-1}). Courtesy of the Fiji Meteorological Service.

TABLE 3.1. Size classification for tropical cyclones based on radius in degrees, measured from the storm centre to the outermost closed isobar.

Radius in degrees	Tropical cyclone size
<2	Very small (midget)
2–3	Small
3–6	Medium to average
6–8	Large
>8	Very large

3.2 Cloud Patterns

The area of clouds associated with tropical cyclones is much larger than the region influenced by strong winds. This is because the strong winds reflect the area characterised by the steep atmospheric pressure gradient, whereas the

FIG. 3.2. Satellite image of Tropical Cyclone Zoe on 27 December 2002, northeast of the main islands of Vanuatu. Note the plumes of high cirrus clouds at the outer storm extremities and the spiralling arms. Base image courtesy of NOAA.

clouds form wherever there is convection associated with the cyclone system, driven by the expanse of warm sea-surface temperatures. As a result, the cloud pattern of a tropical cyclone extends well beyond the rings of close concentric isobars shown on a weather chart, and visible and infrared satellite images are the best way of observing the arrangement of clouds (Fig. 3.2).

At their outer-most extremities, tropical cyclones are marked by high cirrus clouds. Further inwards an even layer of cirrostratus is encountered. Because convection in tropical cyclones is organised into elongated bands that are oriented in the same direction as the horizontal wind, an overhead view of cyclone cloud structure shows the familiar spiral arms that converge inwards. Towards the centre, the convective bands are tightly coiled. Near to the cyclone centre, a dense wall of cumulonimbus towers exists, causing thunderstorm activity with torrential rain and frequent lightning. The round zone of complete cloud cover is known as the *central dense overcast* area, or CDO. Near-circular CDOs are indicative of minimal vertical shear in the atmosphere, a condition that is favourable for the maintenance of tropical cyclone structure. The CDO contains a distinct well-centred eye, which is the hub of storm rotation.

3.3 Eye of the Storm

A distinctive feature of tropical cyclones is their small, comparatively tranquil, central area of fair weather, around which the storm rotates. This is known as the *eye*, which usually measures 20–40 km across (Fig. 3.3). The vivid picture of a small quiet area surrounded by a violent tempest so captures our imagination

FIG. 3.3. Tropical Cyclone Eseta passing near the islands of Tonga on 13 March 2003. The eyewall area of dense cumulus cloud contrasts markedly with the central, almost cloudless eye. Base image courtesy of NASA.

that the phrase 'eye of the storm' has passed into metaphorical usage in everyday language. The reality of the cyclone eye is that any point of land or sea over which it passes experiences extreme changes in weather conditions in a very short span of time. As a cyclone approaches a particular location, the winds build up gradually from gale to storm to hurricane strength,[1] but when the eye arrives the winds die down quickly and almost completely.

The period of calm is short-lived, usually less than an hour, depending on the size of the eye, the speed of motion of the cyclone along its track and whether a particular place is traversed by the full diameter of the eye or only a chord. Then within only minutes as the eye passes away, the winds regain their ferocious hurricane force, although now coming from the opposite quadrant. For people without a proper understanding of this phenomenon, errors of judgement in thinking that the calm eye heralds the end of the storm might clearly have disastrous consequences.

Rainfall, normally falling at high intensity beneath the dense cumulonimbus cloud mass of the *eyewall* (the area immediately bordering the eye) also ceases in the cloudless eye of the storm. There is an accompanying rise in temperature of a few degrees and a drop in relative humidity. It used to be thought that the rise in temperature in the eye was associated with the brief period of sunshine as the eye crosses overhead. This is now known not to be the case, because temperature rises have also been recorded in the eye at night. The slight rise in temperature in the eye is made more noticeable by its contrast with the cooling effect at ground level outside the eyewall, caused by the torrential rainfall and windy conditions.

The fact that the eye is warm is in contrast to other types of tropical depressions that have a cold core of showery activity. The formation of the eye structure is not yet fully understood, but its warm core probably develops through the action of cumulonimbus cloud towers releasing large amounts of latent heat through condensation in the eyewall, and by the adiabatic warming associated with a gentle descent of air in the eye itself. It is this downward vertical current that is responsible for the disappearance of clouds and the brief spell of clear weather as the eye passes overhead.

3.4 Naming Tropical Cyclones

Naming tropical cyclones gives them an identity which is needed to make the public aware of, and prepare them for, the approaching storm (Simpson and Riehl 1981). The Regional Specialized Meteorological Centre (RSMC) in Nadi, Fiji and the other five RSMCs monitoring other cyclone regions around the world, employ the *Dvorak Method* (after its introduction by Dvorak), as the primary technique to determine tropical cyclone intensity.

---

[1] These terms for tropical cyclone strength, based on wind speed, are explained in Chap. 4.

This information is then used to decide when to upgrade the status of a tropical depression by naming it as a tropical cyclone. The Dvorak Method is based mainly upon the interpretation of cloud patterns in visible and infrared images from both polar-orbiting and geostationary satellites (JTWC 2006), and has endured as a reliable technique for over 30 years (Velden *et al.* 2006). Detailed procedures are given for evaluating the satellite signature of a tropical cyclone in terms of its current intensity and predicted near-future intensity (Dvorak 1984). Any other available data sets or observations are also considered, for example from weather radar or remote automatic weather stations sending their readings by telemetry. The Dvorak technique results in a decimal number, called a T-number, that in turn corresponds to an estimate of cyclone intensity (Table 3.2).

It is the usual practice for a tropical depression to be named as a tropical cyclone when two sets of conditions are fulfilled. First, Dvorak intensity analysis should indicate the presence of gale force winds (above 63 km h^{-1}) near the centre of the storm. Second, observations of the recent history and behaviour of the system should suggest that these winds are likely to continue or to strengthen further, rather than die down (Murnane 2004, WMO 2004). Similarly, in the decay stage, a tropical cyclone is downgraded to tropical depression status whenever observations and intensity analysis indicate that the winds near the centre of the system have dropped to less than gale force.

The South Pacific region began assigning female names to cyclones in 1964, and both male and female names have been used from the 1974–1975 cyclone season onwards. The currently designated names for tropical cyclones originating in the RSMC-Nadi area of responsibility are given in

TABLE 3.2. T-number determined by the Dvorak Method and equivalent estimate of tropical cyclone intensity, expressed as wind speed.

T-number	Estimated intensity	
	knots	km h^{-1}
1.0–1.5	25	46
2.0	30	56
2.5	35	65
3.0	45	83
3.5	55	102
4.0	65	120
4.5	77	143
5.0	90	167
5.5	102	189
6.0	115	213
6.5	127	235
7.0	140	259
7.5	155	287
8.0	170	314

1 knot = 1.852 km h^{-1}

TABLE 3.3. Current names for tropical cyclones in the South Pacific, east of 160°E (RSMC-Nadi area of responsibility).

List A	List B	List C	List D	List E (standby)
Ana	Arthur	Atu	Amos	Alvin
Bina	Becky	Bune	Bart	Bela
Cody	Cliff	Cyril	Colin	Chip
Dovi	Daman	Daphne	Donna	Denia
Eva	Elisa	Evan	Ella	Eden
Fili	Funa	Freda	Frank	Fotu
Gina	Gene	Garry	Gita	Glen
Hagar	Hettie	Haley	Hali	Hart
Irene	Innis	Ian	Iris	Isa
Judy	Joni	June	Jo	Julie
Kerry	Ken	Koko	Kala	Kevin
Lola	Lin	Lusi	Leo	Louise
Meena	Mick	Mike	Mona	Mal
Nancy	Nisha	Nute	Neil	Nat
Olaf	Oli	Odile	Oma	Olo
Percy	Pat	Pam	Pami	Pita
Rae	Rene	Reuben	Rita	Rex
Sheila	Sarah	Solo	Sarai	Suki
Tam	Tomas	Tuni	Tino	Troy
Urmil	–	Ula	–	–
Vaianu	Vania	Victor	Vicky	Vanessa
Wati	Wilma	Winston	Wiki	Wano
Xavier	–	–	–	–
Yani	Yasi	Yalo	Yolanda	Yvonne
Zita	Zaka	Zena	Zazu	Zidane

Source: WMO (2005).

Table 3.3. Storms forming west of 160°E in the Coral Sea are named by the Tropical Cyclone Warning Centre in Brisbane, Australia.

The name of a newly formed cyclone is selected in alphabetical sequence by cycling through lists A, B, C and D. The lists are used sequentially from A to D, and the names are chosen continuously through the cycle without reverting to 'A' at the start of each year. Thus, the name of the first tropical cyclone in any given calendar year is the one following alphabetically after the last cyclone name from the previous year, regardless of the position in the current list. In this way, names with initials of all the letters from A to Z are used. Once the last name at the end of List D has been used (Zazu), the series then starts again from the top of List A (Ana), in other words the lists are circular. Names from the standby List E are used as replacements for retired names when necessary. A name is normally retired from the active lists if the storm bearing that name gains notoriety, when for example, it results in a catastrophe for a certain location and is therefore remembered as a legendary event. The retired name is replaced with another name beginning with the same letter and of the same gender. An example is the name 'Heta', which was retired from List A in the current table of names and replaced with 'Hagar', after Tropical Cyclone Heta devastated the single island nation of Niue in January 2004.

Chapter 4
Tropical Cyclone Patterns and Behaviour

4.1 Numbers, Timing and Seasonality

Climatic control of the frequency of tropical cyclones remains poorly understood at the global scale. This means that it is still a mystery why the total number of tropical cyclones, hurricanes and typhoons that develop over the world's oceans each year is about 80 (Table 4.1), or indeed why this number is not much greater or much less (Emanuel 2004). The first of these two observations can be transferred from the global to the regional scale, because at present it is not known why the number of tropical cyclones forming in the South Pacific is about 11% of the world's total. The second point, however, does not hold for the South Pacific. Based on the 1970–2006 climatic record (Appendix 1), although 9 storms is the long-term mean of annual tropical cyclone occurrence, the actual number forming in individual years ranges widely between a minimum of 3 and a maximum of 17 storms (Figs. 4.1 and 4.2).

The year to year variability in tropical cyclone numbers in the South Pacific is mostly related to non-seasonal fluctuations in the Pacific ocean–atmosphere system. In particular, the El Niño-Southern Oscillation (ENSO) phenomenon exerts a major influence, such that tropical cyclone activity is 28% above average for strong Southern Oscillations (Basher and Zheng 1995).[1] There is also a slight tendency in El Niño periods for extra-seasonal tropical cyclogenesis (Hastings 1990). For example, during the 1997–1998 El Niño episode, Tropical Cyclone Keli formed in June 1997, well beyond the end of the normal November to April tropical cyclone season for the South Pacific.

The majority of tropical cyclones occur within a distinct period of the year, so this is widely known as the *cyclone season*. The South Pacific cyclone season coincides with the hot and wet summer season, and traditionally lasts for 6 months from November to April (Fig. 4.3). January and February, at the

---

[1] This figure may be an underestimate as it does not include the 1997–1998 El Niño event.

TABLE 4.1. Average yearly frequency of tropical cyclones according to cyclone region, from 1958 to 1978.

Cyclone region	Frequency	Percentage
Northern hemisphere		
Northwest Atlantic	8.8	11.1
Northeast Pacific	13.4	16.9
Northwest Pacific	26.3	33.2
North Indian Ocean	6.4	8.1
Total	54.6	69
Southern hemisphere		
Northern Australia	10.3	13.0
Southwest Pacific[a]	5.9	7.5
South Indian Ocean	8.4	10.6
Total	24.5	31
Global total	79.1	100

From Gray (1979).
[a]East of longitude 160°E only. Storms formed in the Coral Sea are included here in the total for northern Australia.

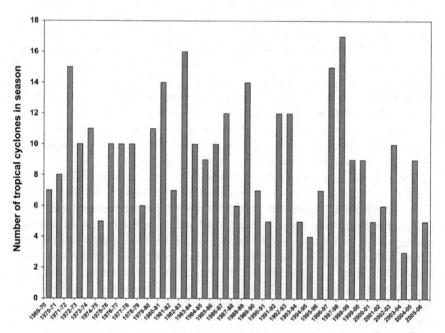

FIG. 4.1. Temporal pattern in the numbers of tropical cyclones in the South Pacific, arranged in cyclone seasons from 1969–70 to 2005–06.

height of the Southern Hemisphere summer, are the 2 months when tropical cyclones are most likely to form (Table 4.2). This is because ocean-surface temperatures in the southwest Pacific warm pool reach their maximum at this time of year, and at the same time the South Pacific Convergence Zone (SPCZ) descends to its most southerly annual position, where it is often well

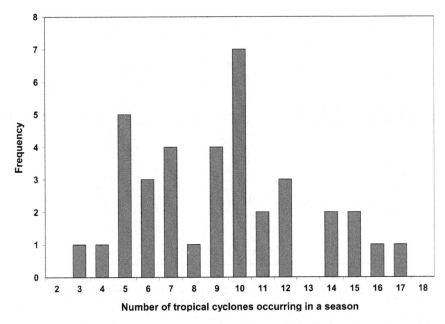

FIG. 4.2. Frequency distribution of the numbers of tropical cyclones occurring per season in the South Pacific, based on the climatic record of all storms from 1970 to 2006.

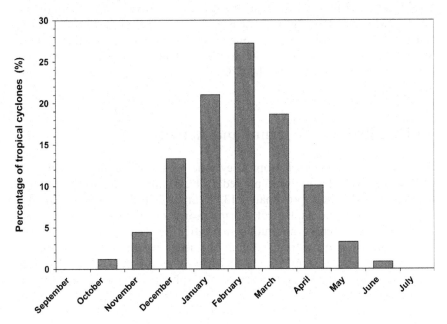

FIG. 4.3. Timing of tropical cyclones in the South Pacific, according to the month within which they were formed, based on the climatic record of all storms from 1970 to 2006.

TABLE 4.2. Historical and recent monthly distributions of tropical cyclone activity in the South Pacific, expressed as percentages of all events (see also Fig. 4.3).

Month	Percentage of recorded tropical cyclones 1789–1923[a]	Percentage of all tropical cyclones 1970–2006[b]
September	0.8	0
October	1.6	0.9
November	3.2	4.0
December	12.4	11.2
January	27.6	19.7
February	19.2	26.7
March	25.6	22.1
April	7.2	11.4
May	0.8	3.3
June	0.8	0.7
July	0.4	0
August	0.4	0
Total	100	100

[a]$n = 250$, longitude 160°E to 140°W only. Source: Visher (1925).
[b]$n = 430$, but note that individual cyclones surviving into consecutive months, e.g. January to February, are counted in both months. Source: Fiji Meteorological Service (2003).

developed as a broad band of active convergence and convection. Since the beginning and end of hot and wet conditions marking the summer season are naturally variable on an inter-annual basis, a small number of cyclones also develop just before and after the traditional hot season, in the months of October, May and June. However, these extra-seasonal storms are often weak and short-lived. No tropical cyclones form in the cooler months of July, August and September during the Southern Hemisphere winter, when sea-surface temperatures drop and the SPCZ migrates equatorwards and is more poorly defined.

4.2 Distribution of Origins and Activity

The median position of all tropical cyclone points of origin from 1939 to 1979 in the South Pacific were plotted by Revell and Goulter (1986a), and was determined to be approximately 13.5°S and 170°E. For average years the spatial pattern in the number of storms shows a distinct trend, decreasing from west to east across the southwest Pacific (Fig. 4.4). This is because the primary influence on tropical cyclone incidence to the west of longitude 170°E is ocean-surface temperature, which becomes progressively cooler from west to east. In comparison, favourable atmospheric conditions, and how far these conditions extend eastwards, are more important than ocean temperature for tropical cyclone activity east of longitude 170°E (Basher and Zheng 1995). On an intra-seasonal basis, the location and migration of the SPCZ is another primary control on tropical cyclone origins.

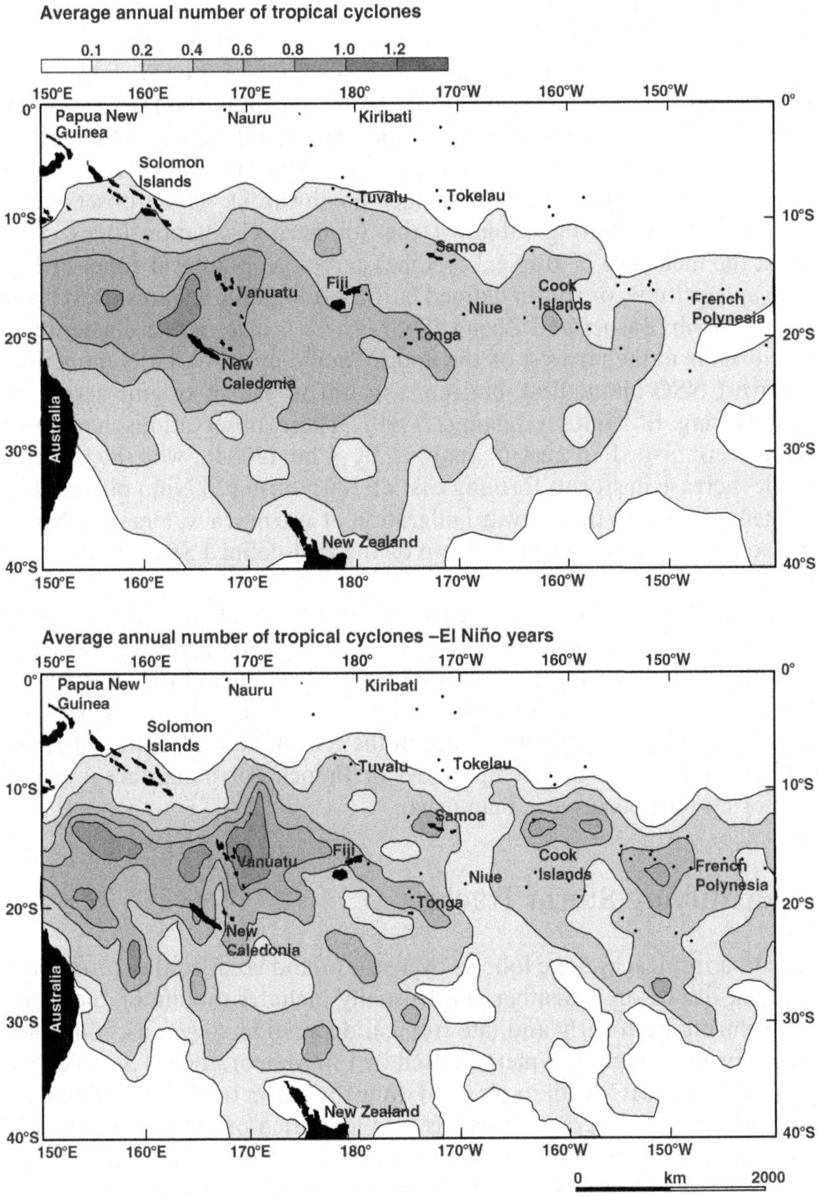

FIG. 4.4. Geographical pattern of tropical cyclone occurrence (risk) in average years and El Niño years. Based on original maps from the Australian Bureau of Meteorology (2005a).

During strong El Niño events, tropical cyclone points of origin undergo substantial changes from the normal pattern, and cyclone activity spreads eastwards across the South Pacific, at least as far as 140°W. Early work quickly determined that fewer cyclones tend to form in the far west of the southwest Pacific region in El Niño periods, but that there is a small increase in the far east. There is also a definite tendency for more storms to develop at lower than median latitudes (farther northwards). These observations were made by comparing values of the Southern Oscillation Index (SOI) against the incidence of tropical cyclones across geographical zones of longitude and latitude, or within defined latitude–longitude squares (Revell and Goulter 1986b, Basher and Zheng 1995). There is also tighter clustering of storm origins in the far west of the South Pacific in strong La Niña periods (negative ENSO anomalies), but more spatial spread of origins across the region during El Niños (Hastings 1990). These observations have been largely reconfirmed in recent analysis by Chu (2004), who identifies a notable increase in storms forming east of 180° during El Niño phases. This is related not only to the eastward migration of warm sea waters in El Niños, but also to a corresponding extension of the well-defined SPCZ in the same direction.

The overall picture that therefore emerges is of tropical cyclones originating mainly to the west of longitude 180° during normal and La Niña years. This means that Solomon Islands, New Caledonia, Vanuatu and Fiji experience a relatively high risk of tropical cyclones compared to nations farther east. During El Niño phases, island groups in the central South Pacific, especially the Cook Islands and the archipelagoes of French Polynesia, face a greater threat of cyclone occurrence than usual.

4.3 Identifying Storm Tracks

The path a tropical cyclone follows over sea or land is called the *storm track*. The track drawn on a weather chart usually includes the life of the storm system during both early and late tropical depression stages, as well as the tropical cyclone stage. The method used by Fiji Meteorological Service forecasters at the RSMC-Nadi to plot and draw the storm track is as follows. An initial area of tropical disturbance is first identified on satellite imagery. If the disturbance exists for 24 h, it is assigned an identification number and tracking begins. The position of the storm centre is observed frequently, and every 6 h its current position is plotted, at 12 midnight, 6 a.m., 12 midday and 6 p.m. The 6-hourly plots on a synoptic map are called *fixes*.

The track continues to be drawn by connecting the fixes, as long as satellite imagery shows that the thermal structure of the system is characterised by a warm core at its centre. If a storm moves south of 25°S, so becoming an extra-tropical system, then the Tropical Cyclone Warning Centre (TCWC) in Wellington, New Zealand, assumes the responsibility of monitoring its position.

Similarly, the TCWC in Brisbane, Australia, takes over the tracking of cyclones that move west of 160°E. Tracking by either the RMSC-Nadi, or whichever TCWC is responsible, ceases when the warm core dissipates, or if a cyclone that has entered subtropical waters takes on structural characteristics that are more typical of a mid-latitude depression.

4.4 Environmental Steering

Nascent tropical disturbances in the South Pacific may stay more or less stationary while developing. Once formed, the majority of tropical cyclones migrate generally eastwards and polewards from their original position. This means that in spite of being relatively small features at synoptic scales, cyclones may cause great damage on any islands they encounter, throughout a belt that is several hundred kilometres wide and more than a thousand kilometres long. Yet predicting the movement of an individual storm remains one of the most difficult tasks in tropical forecasting (McGregor and Nieuwolt 1998). This is because tropical cyclones are notorious for displaying erratic behaviour in their speed and direction.

Environmental steering is the primary influence on the motion of a tropical cyclone. A commonly used analogy is to liken the motion to a leaf being steered by the currents in a stream, except that in the atmosphere the 'stream' has no set boundaries. While an incipient cyclone system is still forming within the SPCZ, and does not yet possess a tall vertical structure, its motion has a tendency to be more erratic compared to later stages of maturity. This is because a young system is steered mainly by surface layer winds, and these winds often show irregular patterns over relatively short distances, owing to the uneven nature of convergence and disturbance of air masses within the SPCZ.

Once a tropical cyclone attains a mature vertical structure reaching high into the tropopause, and if it manages to break free of the influence of the SPCZ by moving southwards, then steering is controlled by *atmospheric steering winds*. The important steering winds are those that occur at the 850–500 hPa level in the mid-troposphere, which are referred to as the *deep layer winds*. The mid-tropospheric wind field tends to be well defined over large areas of the South Pacific. This makes the job of track forecasting somewhat easier through the middle and latter phases of a storm's life cycle (Fiji Meteorological Service 2005, pers. commun.). The example of Tropical Cyclone Tam in January 2006 (Fig. 4.5) illustrates the environmental steering effect. As a result of steering by the mid-tropospheric wind field (the mean deep layer shown in Fig. 4.6), TC Tam tracked in a southeasterly direction between the islands of Fiji and Samoa.

Synoptic-scale influences have important effects on the environmental steering currents. A ridge of high pressure may keep a tropical cyclone entrenched in the trade wind belt, whereas an advancing trough of low pressure can cause a storm track to *recurve*. Recurvature is the term given to

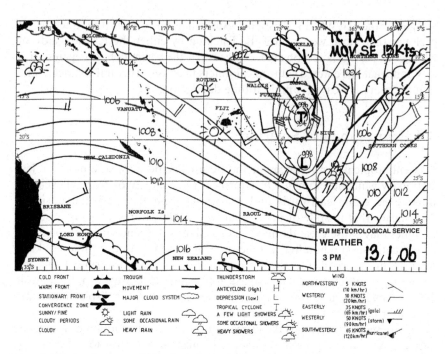

FIG. 4.5. Isobaric chart of Tropical Cyclone Tam at 6 a.m. on 13 January 2006, Fiji Standard Time. Courtesy of the Fiji Meteorological Service.

a deflection in track direction. In tropical latitudes above 15°S, cyclones often move towards the west with only a slight poleward component. This is because of the existence of the subtropical high-pressure ridge farther south. The *subtropical high* is a large anticyclone situated in the subtropics. It is elongated along a west to east axis, thus forming a ridge of high pressure. It is a persistent feature and rarely absent. On the equatorward side of the ridge, easterly winds prevail, pushing cyclones westwards. In the absence of strong steering currents or if the subtropical ridge is weak, then Coriolis acceleration causes storms to turn in a more poleward direction. On the poleward side of the subtropical ridge, westerly winds steer the path of tropical cyclones back towards the east.

In rare cases of two tropical cyclones either forming adjacently or moving close to one another, the steering flow is affected by their interaction. The interaction is subject to the *Fujiwhara effect*, resulting in some atypical behaviour as the two storms swirl towards each other. The Fujiwhara effect is when two vortices are mutually attracted, and as a result have a tendency to spiral around a central point, sometimes merging together. If the two vortices are of unequal size, the larger vortex will tend to dominate the interaction, and the smaller vortex will orbit around it. Therefore, if two cyclones approach each other, their vortices begin interacting in this way. Tropical Cyclone Nancy underwent a minor Fujiwhara effect with Tropical Cyclone Olaf in February 2005 (Fig. 4.7).

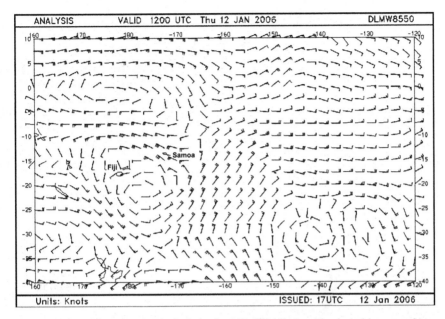

FIG. 4.6. Chart of mean deep winds at the 850–500 hPa level in the mid-troposphere, over the South Pacific at 5 a.m. Fiji Standard Time on 13 January 2006 (5 p.m. GMT on 12 January 2006), during the early life of Tropical Cyclone Tam (shown in Fig. 4.5). Wind feathers are drawn to illustrate the direction from which the wind blows: short barb – 5 knots, long barb – 10 knots, solid triangular barb – 50 knots (1 knot = 1.852 km h^{-1}). Note the northwesterly wind-field pattern south of 10°S and east of 180°, in the vicinity of Samoa. As a result of steering by this wind field, TC Tam followed a southeasterly course in this region. Courtesy of the Australian Bureau of Meteorology.

4.5 Speed of Advance

Tropical cyclones advance along their track at slow speeds compared to the extremely fast winds (described in Chap. 5) revolving around the storm centre. The speed of cyclone movement is usually under 25 km h^{-1}, equivalent to 600 km or less per day. Yet there is great variability in speed, both between separate cyclones and also during the lifespan of an individual system. As a general rule, a cyclone vortex moves more slowly in the early phase of its life, but then gains speed upon reaching maturity. Tropical cyclones also have a habit of displaying erratic and unpredictable changes in their rate of movement, especially while occupying lower latitudes. They often show a tendency to decelerate, especially if they veer in unusual directions along inflections or looping tracks. Some tropical cyclones stop dead in their tracks and remain almost stationary for a period of several hours, before accelerating once again.

Many cyclones lose their strength and dissipate before leaving tropical latitudes, but those that do survive as they migrate into subtropical and mid-latitudes tend to experience significant acceleration. Vigorous extra-tropical

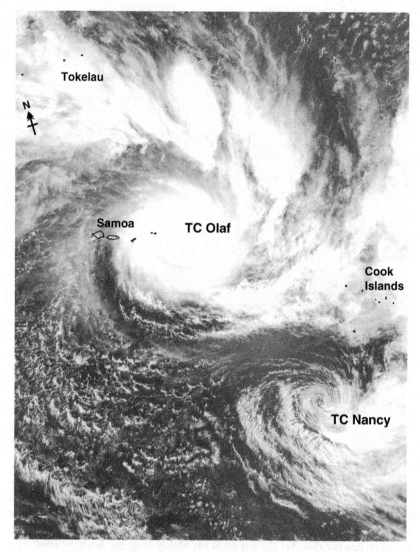

FIG. 4.7. Adjacent Tropical Cyclones Olaf and Nancy on 16 February 2005, interfering with each other's movement. Interference between mutually attracted vortices is known as the Fujiwhara effect. Base image courtesy of NASA.

cyclones often travel most rapidly, sometimes reaching double the velocity they had earlier while occupying tropical waters.

Figure 4.8 demonstrates a fairly typical story of several changes in the speed of a tropical cyclone, in this case during the progress of TC Gavin near Tuvalu and through Fiji in early March 1997. As the system developed to the far west of Tuvalu from 2 to 4 March it travelled approximately 450 km in 48 h on an easterly track, from longitude 172°E to 176°E, at an average speed

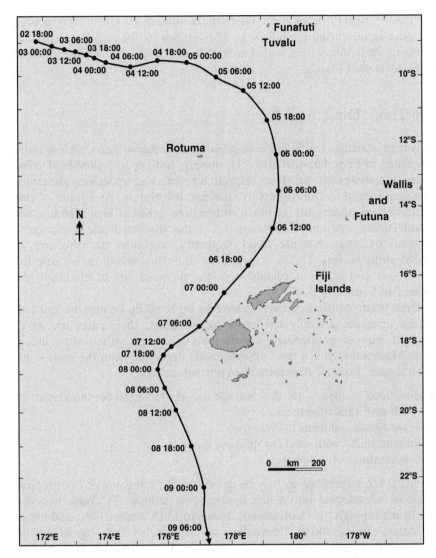

FIG. 4.8. Track of Tropical Cyclone Gavin through Fiji waters in March 1997. Fixes in position of the centre of the eye at 6-hourly intervals in GMT times (dd hh:mm), are shown by the black markers. Changing distances between successive markers illustrate acceleration or deceleration in the speed of the cyclone along its path.

of 9.4 km h⁻¹. Thereafter, TC Gavin accelerated to a steady 20 km h⁻¹ as the storm turned first southeast then south, passing to the west of Wallis and Futuna on 6 March. As TC Gavin approached and subsequently passed the Fiji Islands on a southwesterly course, its speed increased further to 22 km h⁻¹. In contrast, over the next 18 h from 12 midday on 7 March until 6 a.m. on 8 March, the system decelerated rapidly and advanced slowly at an average of

8.5 km h^{-1}, with the track recurving to the southeast. TC Gavin then speeded up once again during the following 24 h, shown by the increasing distance between the 6-hourly fixes, and continued to gain pace while leaving southern Fiji waters on 9 March.

4.6 Track Directions

Any map of tropical cyclone tracks covering a number of years, such as those presented in Figs. 4.9, 4.10 and 4.11, appears initially as a jumble of criss-crossing storm paths. Yet closer inspection reveals that underlying the tangle is a major trend in a northwest to southeast orientation. As a result of this trend in track alignment, the north and western flanks of high islands in the South Pacific commonly find themselves on the 'windward' side in relation to tropical cyclones, because they frequently face into the direction of approaching storms. This is a reversal of the usual situation, because the north and west coasts of islands lie on the sheltered side in relation to the dominant Southeast Trade Winds.

Even where tropical cyclones behave as expected by keeping to 'normal' tracks oriented generally northwest to southeast, their paths are never straight. Any curves often have a gentle parabolic shape, without sharp inflection. Many other storm tracks display wide departure from the more common shapes. Types of departure from normal include:

- directions contrary to the average trend, for example northeast to southwest track alignment,
- sharp twists and turns in direction,
- looping tracks with single or multiple loops,
- combinations of the above.

Unusual track orientations may be caused by the remnants of one tropical cyclone redeveloping into a new system. For example, TC Wasa travelled southeast through the Cook Islands from 6 to 13 December 1991, and began to decay east of 140°W. Thereafter, the remains of the original system restrengthened unexpectedly and formed into TC Arthur, which steered an anomalous northeast course through eastern French Polynesia from 14 to 18 December (Fig. 4.12).

In cases of looping storm tracks, the loops may display either tight or open curvature. Loops can be followed in anticlockwise or clockwise directions. More rarely, double loops are observed (Fig. 4.13). All of these variations show that although an average picture of track shape, curvature and orientation can be reasonably defined for tropical cyclones in the South Pacific, nothing of the sort can be relied on for individual storms.

Since the advent of geostationary satellites and the images they provide, it has been noted that the position of a cyclone eye sometimes oscillates, relative to the apparent focus of storm rotation. This curious motion is known as *wobble* and is probably caused by asymmetrical convection in the warm inner

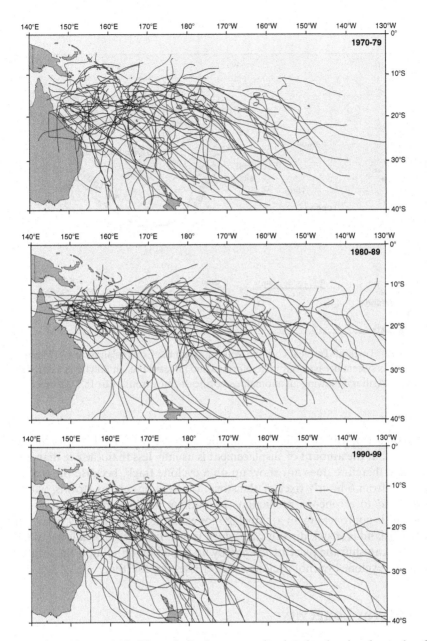

FIGS. 4.9., 4.10. AND 4.11. Three charts for consecutive decades showing the tracks of all tropical cyclones in the South Pacific, over the 30-year period from 1970 to 1999.[2] Note the decrease in the number of tracks from west to east on each chart, and the overall northwest to southeast trend underlying the crisscrossing pattern. Plotted from original data provided by the Fiji Meteorological Service.

[2] This is the best-track data available from the Fiji Meteorological Service. Original FMS source data are archived in digitised GIS format and are more detailed and accurate than track images available elsewhere. The accuracy of some storm tracks in the early 1970s may be less reliable, during the period of emerging satellite technology. The tracks include the lifespan of both incipient and decaying systems during their tropical depression stages.

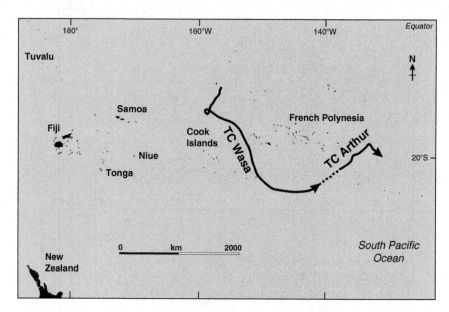

FIG. 4.12. Track of Tropical Cyclone Wasa through the Cook Islands, 6–13 December 1991. This system decayed but then reformed and affected French Polynesia as Tropical Cyclone Arthur, following an anomalous northeast track from 14 to 18 December.

core of a tropical cyclone. Wobbling occurs over periods of approximately 6–10 h, but the amount of displacement is usually less than the eye diameter. Wobble therefore does not show up on a cyclone track, because the track is plotted from 6-hourly fixes of the storm centre. For current storms, wobble may lead to erroneous predictions of cyclone motion and direction. If a cyclone is approaching an archipelago for example, wobbling may determine which island is struck directly and which is not. In the case of a large island, wobble can shift the place of *landfall* by up to 80 km. Landfall is the location where a cyclone eye (not cyclone edge) crosses the coastline.

4.7 Storm Longevity

For the purpose of this section, the lifespan refers to the period that a storm remains classified as a tropical cyclone, based on its structure and intensity as described in Chap. 3. This is independent of position, so includes any time that a cyclone spends in extra-tropical waters, for those storms that migrate beyond latitude 25°S. Tropical and extra-tropical depression phases at the start or end of a cyclone's life are not included. Under this system, the modal lifespan of tropical cyclones in the South Pacific is 4 days, with 67% of all storms lasting between 3 and 8 days (Fig. 4.14). Exceptionally long-lived

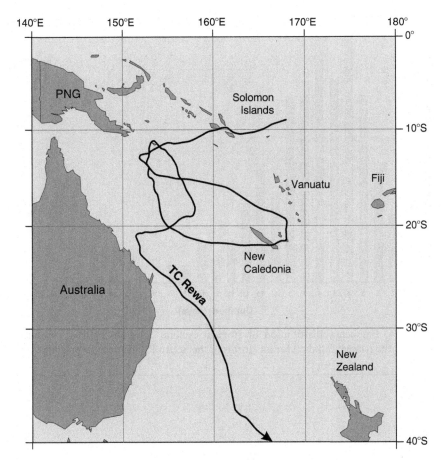

FIG. 4.13. Track of Tropical Cyclone Rewa, from 28 December 1993 to 23 January 1994. This storm lasted over 3 weeks, following a highly convoluted, double-looping path. It affected parts of Solomon Islands, Papua New Guinea (twice), New Caledonia, Vanuatu and Australia with strong winds and heavy rainfall.

tropical cyclones may survive for more than 3 weeks. In most cases, these exceptions are storms that travelled slowly or followed peculiarly sinuous tracks. The sinuosity caused them to remain in tropical latitudes over warm sea surfaces, hence maintaining their intensity for extended periods.

For example, TC Rewa (Fig. 4.13) lasted from late December 1993 through most of January 1994. The total lifespan of 26 days is the longest cyclone duration on record. The storm followed a highly convoluted track with a rare double loop. This kept the centre of the storm located over the warm Coral Sea for much of its life. Even as TC Rewa began moving south, it travelled down the east coast of Australia, remaining over the warm East Australian Current. These warm ocean conditions refuelled the system, prolonging its longevity many days beyond the average lifespan.

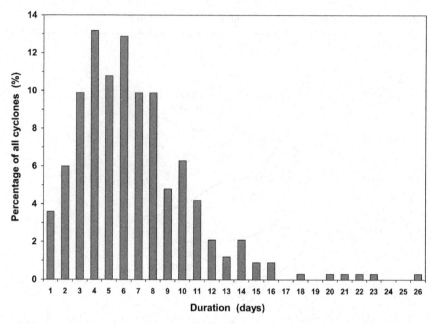

FIG. 4.14. Frequency distribution of tropical cyclone durations in days (rounded down to the nearest full day) for all storms in the South Pacific over 1970–2006.

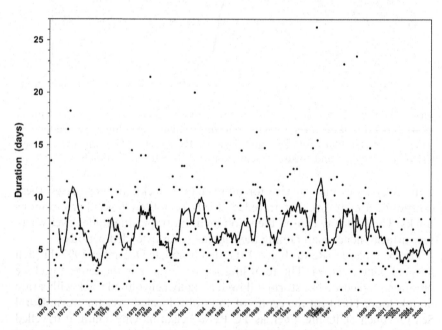

FIG. 4.15. Longevity of individual tropical cyclones (black markers) from 1970 to 2006, showing the running mean of nine storms (solid line). Note that the markers are spaced equidistantly along the horizontal, so that the time axis is non-linear.

TABLE 4.3. Summary of selected tropical cyclone characteristics in the South Pacific, 1970–2006.

Feature	Details	Notes
Seasonality		
Cyclone season	November to April	
Mean occurrence	9.2 cyclones	
Maximum occurrence	17 cyclones, 1997–1998 season	El Niño episode
Minimum occurrence	3 cyclones, 2003–2004 season	
Month with highest Frequency	February, 27% of all storms	
Earliest cyclone	9–12 October 1997 (TC Lusi)	El Niño episode
Latest cyclone	7–17 June 1997 (TC Keli)	El Niño episode
Origins and movement		
Median point of origin	14°S 170°E	For 1939–1979 only (Revell and Goulter 1986a)
Average track orientation	NW–SE	
Storm longevity		
Mean	7 days	
Mode	4 days	

Through the climatic record from 1970 to 2006, neither an increasing nor a decreasing trend in tropical cyclone longevity is observed (Fig. 4.15). An oscillating pattern of average storm duration through time is seen from a plot of running means of nine consecutive tropical cyclones.[3] The oscillations do not appear to be entirely random, but climatic influences on any periodicity in the pattern have not yet been interpreted. A summary of selected tropical cyclone behavioural characteristics is given in Table 4.3.

[3] Nine cyclones were chosen as the basis for calculating running means, since this is the long-term average frequency per cyclone season.

Chapter 5
Meteorological Conditions

"... the wind rises, with threatening gusts that gather strength; low clouds drive furiously; and lightening and thunder, wind and rain, work up in a quick crescendo to the full blast of the hurricane. Rain drives in horizontal rods that sting and blind so that they cannot be faced. Tall coconut palms lash and bend till their heads sweep the ground or their trunks snap with loud reports. The air seems a solid rushing mass."

(Derrick 1951, p. 115)

5.1 Low Pressure

As a tropical cyclone approaches, the barometer falls, at first slowly and then more quickly until it reaches an alarmingly low level. The pressure gradient at sea level near the centre of a mature cyclone is very steep, as shown by closely spaced isobars on a weather chart (Fig. 5.1). It is the steepness of the pressure gradient that accounts for the high wind velocities. Low pressure also has an important influence on storm surge (Section 5.3). The minimum pressure attained is one of two main parameters used to indicate tropical cyclone intensity. The other parameter is wind strength, discussed in Section 5.2.

For South Pacific tropical cyclones, the lowest pressures average around 975 hPa,[1] and 50% of storms drop to between 955 and 985 hPa. Exceptional storms of phenomenal intensity may fall well below this level. Tropical Cyclone Zoe is the most intense storm on record, and plummeted to a minimum pressure of 890 hPa on 29 December 2002 when it was located just east of Tikopia in the outer Solomon Islands. Over the last three and a half decades, available data (Fig. 5.2) exhibit periodic oscillation in minimum low pressure values, and a slight downward trend that might be related to global warming and associated sea-surface temperature rise (see Chap. 6).

---

[1] The units of hectopascals (hPa) and millibars (mb) for recording barometric pressure are equivalent.

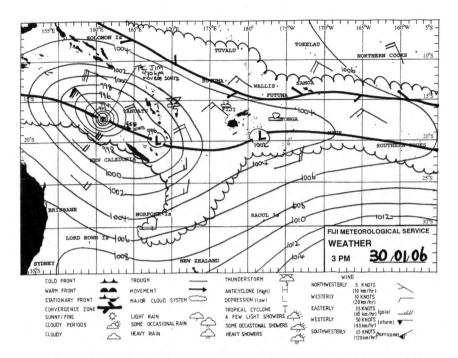

FIG. 5.1. Weather chart showing the closely spaced isobars around Tropical Cyclone Jim, northwest of New Caledonia on 30 January 2006. Courtesy of the Fiji Meteorological Service.

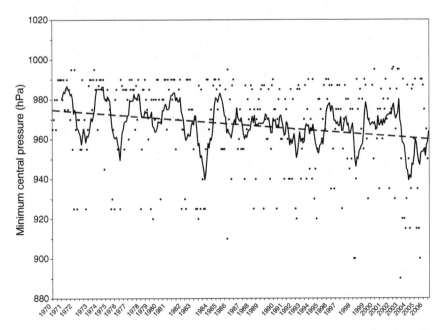

FIG. 5.2. Minimum recorded central pressure of individual tropical cyclones (black markers) from 1970 to 2006, showing the running mean of nine storms (solid line). Note that the markers are spaced equidistantly along the horizontal, so the time axis is non-linear.

5.2 Strong Winds

5.2.1 Wind Effects

It is the steep barometric pressure gradient established between a tropical cyclone's outer edge and its central low that causes strong winds to blow inwards. Violent winds are one of the most dangerous and costly aspects of cyclones for people living on Pacific islands (Table 5.1). Strong winds have a number of effects. They generate large waves at sea and drive these onshore, damaging coral reefs and causing severe coastal erosion. They also contribute to storm surge, resulting in widespread sea flooding of low-lying shorelines. Fierce winds tear at vegetation, stripping foliage, uprooting trees and flattening crops. Large amounts of salty sea spray whipped up from the ocean are blown inland and poison coastal forests. Figures 5.3 and 5.4 illustrate the destructive effects of Tropical Cyclone Heta in January 2004 on the coastal forest of Niue island.

One of the reasons why high wind speeds are very damaging is that for a non-aerodynamic shape such as the flat side of a building, the pressure exerted on the surface facing the wind is roughly proportional to the square of the wind speed. So, for example, 100 km h^{-1} winds will exert four times as much pressure as 50 km h^{-1} winds, not twice as much (Krishna 1984). In addition, on the leeward side of flat obstacles, a partial vacuum is created. This causes an outward suction effect acting in the same direction as the wind pressure on the opposite side of the obstacle. Together these forces can wreak much destruction. Flying debris such as corrugated roofing materials also poses a significant hazard. Figure 5.5 shows a ruined coastal village in eastern Fiji, two days after Tropical Cyclone Ami struck in January 2003, where many people were left homeless.

TABLE 5.1. Maximum wind speeds and costs of resulting damage in the Kingdom of Tonga for tropical cyclones over the 20-year period 1982–2001.

Tropical cyclone	Year	Maximum winds (knots[a])	Cost of damage (million T$[b])	Deaths
Isaac	1982	130	18.7	6
Ofa	1990	140	3.2	1
Sina	1990	100	–	0
Kina	1993	120	–	3
Hina	1997	90	–	0
Ron	1998	125	–	0
Cora	1998	75	19.6	0
Waka	2001	140	104	1

Source: Tonga Meteorological Service.
[a]1 knot = 1.852 km h^{-1}.
[b]T$1 (1 Tongan Pa'anga) is equivalent to 0.38 € and US$0.50 (January 2007 rates, not adjusted).

Fɪɢ. 5.3. Vegetation on the western coastline of Niue island, stripped of foliage by high winds and waves during Tropical Cyclone Heta in January 2004. The surface of the emerged marine terrace, which was overtopped by waves, is approximately 23 m above sea level. Photo by Mosmi Bhim.

Fɪɢ. 5.4. Dead forest on Niue island, poisoned by salty sea spray that was blown inland by Tropical Cyclone Heta in January 2004. Photo by Douglas Clark.

FIG. 5.5. Demolition of a village in the Lau Islands of eastern Fiji by Tropical Cyclone Ami in January 2003. Source: Fiji Navy, supplied by the Fiji Ministry of Information.

5.2.2 Wind Strength

Wind strength is the main parameter used for describing the intensity of a tropical cyclone (Table 5.2), to complement readings of low barometric pressure at the storm centre. The strength of the winds is determined from the average wind speed, referred to as the *sustained wind speed*. Sustained winds are recorded over 10-min averaging times, above ground level in an open and flat space. It should be understood that winds in a tropical cyclone are extremely squally in nature, and gusts much stronger than the sustained winds are frequently experienced over burst of a few seconds.

The strongest winds in a tropical cyclone are termed *hurricane-force* winds. Less severe winds are described as either *storm force* or *gale force*. There is no upper limit given for wind speeds in the hurricane-force category, and tropical cyclone wind speeds can far exceed the lower limit of 64 knots (118 km h^{-1}). This classification for wind strength ties in with the traditional Beaufort Scale[2] that includes descriptions of the visible effects of wind over both land and sea (Table 5.3). It is seen that tropical cyclone winds cover Beaufort categories 8–12, 12 being the maximum category on scale. Nowadays, the top

---

[2] The traditional Beaufort Scale for wind strength continues to be utilised by RMSC-Nadi in the tropical South Pacific region. Wind speed on the Beaufort Scale can be expressed by the formula: $v = 3.0132 \, B^{1.5}$ (km h^{-1}), where v is wind speed and B is the Beaufort number.

TABLE 5.2. Classification of tropical cyclone intensity and wind strength, based on sustained wind speeds measured over 10-min intervals.

Tropical cyclone intensity	Wind strength	Wind speed					
		knots		km h⁻¹		m s⁻¹	
		Average	Gusts	Average	Gusts	Average	Gusts
Gale	Gale force	34–47	Up to 65	63–87	Up to 120	17–24	up to 33
Storm	Storm force	48–63	Up to 85	88–117	Up to 157	25–32	up to 44
Hurricane	Hurricane force	≥64	≥86	≥118	≥158	≥33	≥45

TABLE 5.3. Beaufort Scale of wind force.

Force	Speed (knots)	Speed (km h⁻¹)	Description	Specifications for use on land	Specifications for use at sea
0	0–1	0–1	Calm	Calm; smoke rises vertically	Sea like a mirror
1	1–3	1–5	Light air	Direction of wind shown by smoke drift, but not by wind vanes	Ripples with the appearance of scales are formed, but without foam crests
2	4–6	6–11	Light breeze	Wind felt on face; leaves rustle; ordinary vanes moved by wind	Small wavelets, still short, but more pronounced. Crests have a glassy appearance and do not break
3	7–10	12–19	Gentle breeze	Leaves and small twigs in constant motion; wind extends light flag	Large wavelets. Crests begin to break. Foam of glassy appearance. Perhaps scattered white horses
4	11–16	20–28	Moderate breeze	Raises dust and loose paper; small branches are moved	Small waves, becoming larger; fairly frequent white horses
5	17–21	29–38	Fresh breeze	Small trees in leaf begin to sway; crested wavelets form on inland waters	Moderate waves, taking a more pronounced long form; many white horses are formed. Chance of some spray
6	22–27	39–49	Strong breeze	Large branches in motion; whistling heard in telegraph wires; umbrellas used with difficulty	Large waves begin to form; the white foam crests are more extensive everywhere. Probably some spray
7	28–33	50–61	Near gale	Whole trees in motion; inconvenience felt when walking against the wind	Sea heaps up and white foam from breaking waves begins to be blown in streaks along the direction of the wind

(continued)

TABLE 5.3. (continued)

Force	Speed (knots)	Speed (km h⁻¹)	Description	Specifications for use on land	Specifications for use at sea
8	34–40	62–74	Gale	Breaks twigs off trees; generally impedes progress	Moderately high waves of greater length; edges of crests begin to break into spindrift. The foam is blown in well-marked streaks along the direction of the wind
9	41–47	75–88	Severe gale	Slight structural damage (chimney-pots and slates. removed).	High waves. Dense streaks of foam along the direction of the wind. Crests of waves begin to topple, tumble and roll over. Spray may affect visibility
10	48–55	89–102	Storm	Seldom experienced inland; trees uprooted; considerable structural damage	Very high waves with long overhanging crests. The resulting foam, in great patches, is blown in dense white streaks along the direction of the wind. On the whole the surface of the sea takes on a white appearance. The 'tumbling' of the sea becomes heavy and shock-like. Visibility affected
11	56–63	103–117	Violent storm	Very rarely experienced; accompanied by wide-spread damage	Exceptionally high waves (small and medium-size ships might be for a time lost to view behind the waves). The sea is completely covered with long white patches of foam lying along the direction of the wind. Everywhere the edges of the wave crests are blown into froth. Visibility affected
12	≥64	≥118	Hurricane	Severe and extensive damage	The air is filled with foam and spray. Sea completely white with driving spray; visibility very seriously affected

Categories 8–12 describe gale, storm and hurricane-force winds in tropical cyclones.

end of the Beaufort Scale is extended by the Saffir–Simpson Scale (Simpson and Riehl 1981). This is used to classify hurricane-intensity tropical cyclones into five *hurricane categories*, 1–5 (Table 5.4).

In the North Atlantic Ocean and the Caribbean Sea, the term 'hurricane' is widely used as a synonym for 'tropical cyclone'. Some confusion may be caused by this terminology because not all tropical cyclones have winds that reach hurricane force. Even those cyclones that do eventually strengthen to hurricane intensity at some time during their lifespan are characterised in their developmental phase, and later in their decay stage, by maximum winds of only gale or storm force. To avoid confusion when public warnings are issued, forecasters in the South Pacific therefore refer to a tropical cyclone with maximum winds of gale, storm or hurricane force, depending on the current intensity at the time of observation. Figure 5.6 indicates that nearly half of all tropical cyclones developing in the South Pacific over the 36-year period from 1970 to 2006 attained hurricane intensity.

5.2.3 *Wind Distribution*

Figure 5.7 demonstrates a simplified horizontal distribution of wind strengths within an idealised and stationary tropical cyclone that has reached hurricane intensity. One of the most remarkable features of tropical cyclones compared to other types of storms is that the strongest winds are felt just outside the central eye, which by comparison is a zone of almost complete calm. Farther away from the central zone of hurricane-force winds, the winds decrease in strength. The boundaries of wind strength are approximately circular and concentric, but the ring of hurricane-force winds typically has a much smaller radius than the ring of storm-force winds, that in turn has a smaller radius than the zone experiencing gale-force winds.

The progression of a storm along its track, assisted by the steering winds, contributes to the wind field. This means that in the Southern Hemisphere on the left side of the track, the speed of cyclone movement adds to the speed of the rotating winds, whereas wind speeds on the right side of the track are

TABLE 5.4. Saffir–Simpson Scale for categories of hurricane-force tropical cyclones (NOAA 2006c).

Hurricane category	Maximum sustained winds[a]			Minimum central pressure[b]
	knots	km h^{-1}	m s^{-1}	mb
1	64–82	119–153	33–42	≥980
2	83–95	154–177	43–49	979–965
3	96–113	178–209	50–58	964–945
4	114–135	210–249	59–69	944–920
5	≥136	≥250	≥70	<920

[a]1-min means.
[b]Classification of hurricane category by central pressure ended in the 1990s, and wind speed alone is now used. These estimates of the central low pressure that accompany each category are therefore for reference only.

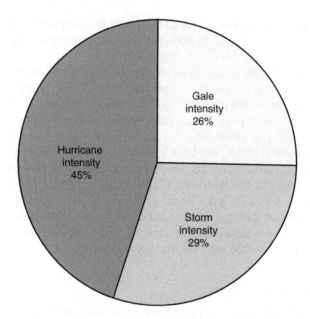

FIG. 5.6. Pie chart illustrating the percentages of tropical cyclones in the South Pacific 1970–2006 that reached gale intensity, storm intensity or hurricane intensity, according to the maximum sustained wind speeds. Based on original data provided by the Fiji Meteorological Service.

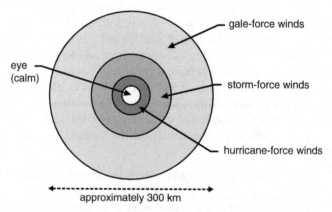

FIG. 5.7. Simplified horizontal zonation of wind strengths in an idealised and stationary tropical cyclone which has reached hurricane intensity. Adapted from Krishna (1984).

reduced by a similar amount (Fig. 5.8). As a consequence, the fiercest winds are felt on the left side of tropical cyclone tracks (Fig. 5.9). This means that places lying equidistant but on opposite sides of a storm track may therefore experience winds of contrasting intensities.

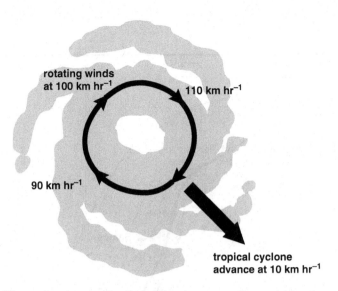

FIG. 5.8. Illustrating the contribution of the movement of a tropical cyclone along its track to the speed of its rotating winds. Adapted from NOAA (2006b).

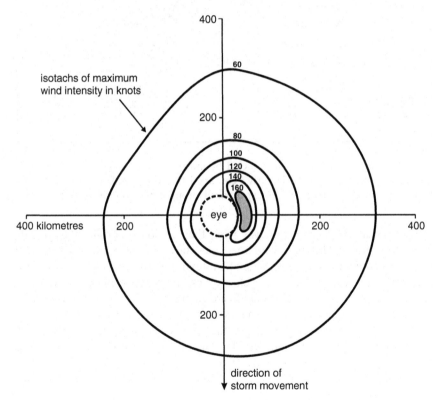

FIG. 5.9. Model of the wind field for a hurricane-intensity tropical cyclone drawn with respect to an arbitrary southward direction of motion. Adapted from Simpson and Riehl (1981).

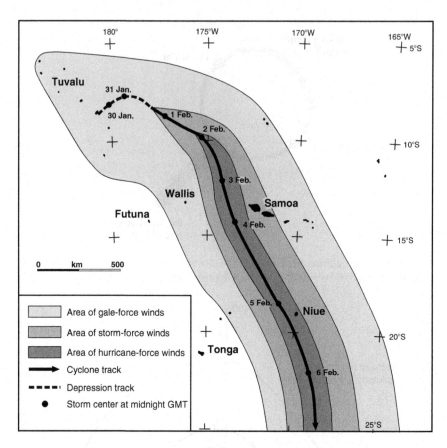

FIG. 5.10. Sea areas and islands affected by hurricane, gale and storm-force winds during Tropical Cyclone Ofa in early February 1990. The islands of Tuvalu, Wallis, Samoa, Niue and northern Tonga experienced strong winds and much damage.

As tropical cyclones travel away from their region of origin, the roughly concentric rings of different wind intensities shown in Fig. 5.7 will be extended into elongated bands on either side of, and parallel, to the cyclone track. This is illustrated in a typical map of cyclone track and intensity in Fig. 5.10, in this case for Tropical Cyclone Ofa that affected a number of island nations in early February 1990. The map shows that Wallis island experienced only gale-force winds, whereas Samoa lay closer to the track and within the boundary of storm-force winds. Niue suffered the full brunt of hurricane-force winds because the track passed less than 50 km to the west of the island.

5.2.4 Wind Direction

Winds drawn in towards the centre of a tropical cyclone do not trace the circular or elliptical shape of the isobars shown on synoptic weather charts, but instead follow spiral streamlines (Fig. 5.11). The angle of deviation of wind

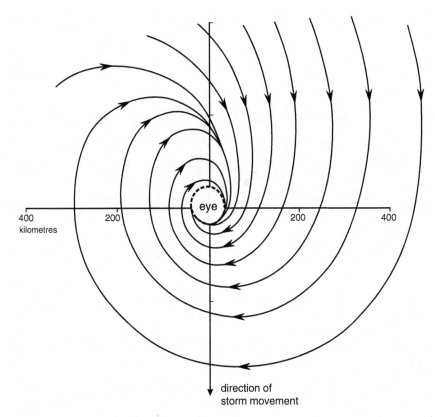

FIG. 5.11. Surface streamlines for an intense tropical cyclone, drawn with respect to an arbitrary southward direction of motion. Adapted from Simpson and Riehl (1981).

bearing from the isobar can be measured, and varies from 20° to 40°, depending on the distance from the storm centre. Localised influences interfere with the general pattern of streamlines. There is much disruption of air flow over and around obstructions, such as irregular topography or buildings in urban areas, giving gusty winds with erratic changes in direction. Squally cells of heavy rain embedded within the storm also tend to show frequent fluctuations in wind pattern.

The combined effect of the inward-spiralling wind pattern and the general advance of a tropical cyclone along its track means that at a particular location the bearing of the wind changes as a storm approaches, passes nearby and then moves away. Such shifts in wind direction are illustrated in Fig. 5.12, which shows the track of Tropical Cyclone Eric in January 1985 across the southern tip of Espiritu Santo island in Vanuatu. At Luganville town on the southeast coast of Santo, the wind first blew from the northeast while the cyclone approached from the west. As TC Eric came nearer, the winds veered round to the northern quadrant, at the same time intensifying from gale to

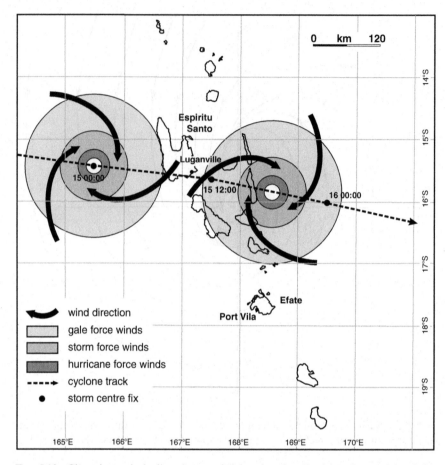

FIG. 5.12. Changing wind direction and intensity for Espiritu Santo island in Vanuatu, during the traverse of Tropical Cyclone Eric from 15 to 16 January 1985. At Luganville town in the path of the eye, the wind bearing swung around a full 180°, veering from the northeast as the cyclone approached to the southwest as it migrated away. Dates and times (dd hh:mm) are given in GMT.

storm to hurricane force. Conditions then dropped to calm as the eye passed over Luganville shortly before midday on 15 January (GMT).

On the other side of the eye, hurricane-force winds returned again, but now blowing from the south. Thereafter the wind intensity waned to storm and then to gale force. The wind bearing veered to the southwest as the track took TC Eric away to the east of Santo island. Overall the wind bearing at Luganville swung round a full 180° from northwest to southeast during the direct overhead traverse of TC Eric. Elsewhere in Vanuatu fluctuations in wind direction were also experienced, but the swing in degrees as the storm passed was less for islands farther away from the line of the track.

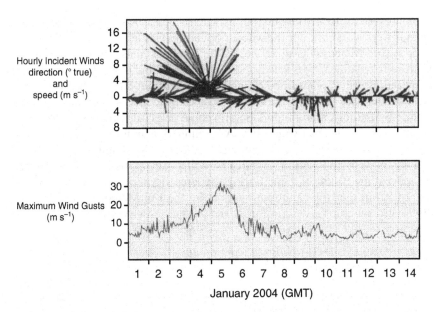

FIG. 5.13. Relationship between shifting wind direction and fluctuating maximum wind speed at Apia in Samoa, during the passage of Tropical Cyclone Heta in early January 2004. Extract from Australian Bureau of Meteorology (2004).

Figure 5.13 provides another example, showing the relationship between fluctuating wind direction and maximum wind speed measured at Apia, the capital of Samoa, during the passage of Tropical Cyclone Heta in early January 2004. Shifting wind directions with respect to the orientation of a coastline, especially whether winds are offshore or onshore, are important for determining the extent of sea flooding by storm surge. This is examined in Section 5.3.

5.2.5 Case Study – Intensity of Tropical Cyclone Ofa, February 1990

Tropical Cyclone Ofa in early February 1990 was a devastating event for several South Pacific nations (Fig. 5.10). The storm initially developed over Tuvalu on 27 January as a depression within the South Pacific Convergence Zone. It rapidly intensified to hurricane strength as it moved south southeast near the Samoan islands with sustained winds greater than 118 km h^{-1}. Samoa was severely affected from 3 to 4 February. Winds became very destructive, lasting for almost 24 h, with rain continuous and heavy for a few hours longer.

The combination of furious winds, heavy rainfall and storm surge created an impact not encountered in Samoa in more than a hundred years. Huge waves and sea spray resulting from the storm tide flooded low-lying areas, reaching up to 10–15 km inland. Northern coasts of Upolu and Savai'i islands were worst hit, but the whole population was left in a state of terrible distress (Fiji Meteorological Service 1990). Roofs of houses were peeled away, walls knocked down, trees felled, and roads, bridges and power lines were badly damaged. Storm surge washed off and reshaped about 80% of the northwestern coastlines of the two main islands, adding to the widespread flooding caused by the torrential rainfall. Seven people perished, either swept away by giant waves or killed by flying debris. According to preliminary estimates of the damage by the media, TC Ofa cost Samoa about US$130 million.

The single island nation of Niue was also devastated. The central eye passed within 50 km of Niue during the late afternoon on 4 February. An excerpt from the tropical cyclone report by the Fiji Meteorological Service (1990) gives a graphic description of the storm's ferocity:

As the eye of Ofa passed close to Niue, destructive hurricane-force winds lashed the island for several hours. Gigantic sea waves resulting from storm surge swept over the northern and western coastal areas of the island and were reported to have reached several metres high. Virtually all landings to the sea were washed away or badly damaged by huge sea waves. There was considerable damage to hospital buildings, the island's hotel, roads, houses, churches, community halls and other facilities for the public. Due to the damage to power lines, electricity was out for about 24 hours. Most of the island's private water supply tanks were contaminated by salt water and declared unsuitable for drinking. Luckily, there was no loss of life or serious injury. The total loss from the cyclone was estimated at around US$2.5 million.

5.3 Storm Surge and Sea Flooding

5.3.1 Wind and Pressure Components

Widespread sea flooding by storm surge around the coastlines of South Pacific islands is a serious hazard during tropical cyclones. For example, on 25 February 2005, Tropical Cyclone Percy, a hurricane-intensity system with sustained winds up to 249 km h^{-1}, had a severe impact on the atoll nation of Tokelau (Fig. 5.14). Tokelau consists of three atolls: Fakaofo, Nukunonu and Atafu. The storm surge generated by TC Percy inundated all three atolls. The high surge allowed the powerful winds to send waves sweeping across the low-lying atoll islands (Fig. 5.15). Waves swept over from both the ocean and the lagoon sides of the atolls, clashing in the middle of the islands and swamping villages (OCHA 2005).

There are two main components to storm surge: violent winds and low atmospheric pressure at sea level (Fig. 5.16). The first of these two components is the major one. As the winds gradually become stronger in a maturing storm, the wind stress acting on the surface of the ocean increases. This generates large swells out at sea and drives giant waves against coastlines (Fig. 5.17). This has the effect of piling the sea up against the shore. The secondary component

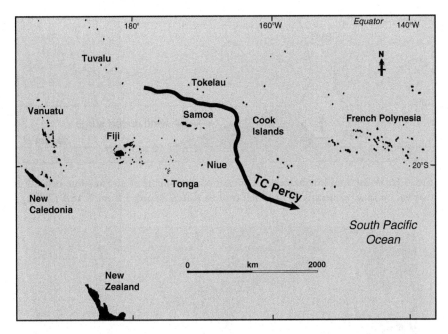

FIG. 5.14. Path of Tropical Cyclone Percy through the atoll nation of Tokelau and the Cook Islands in late February 2006. TC Percy caused sea flooding of several atolls, including Nukunonu (9.1°S, 171.5°W) in Tokelau and Pukapuka (10.5°S, 165.5°W) in the Cooks.

FIG. 5.15. Sea flooding of Nukunonu atoll in Tokelau, during Tropical Cyclone Percy on 26 February 2005. Photo courtesy of the Tokelau Public Works Department.

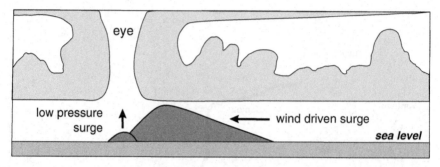

FIG. 5.16. Wind and low barometric pressure components of storm surge beneath a tropical cyclone advancing right to left (not to scale). Adapted from NASA (2006).

FIG. 5.17. Furious seas lashing the coastline of Tikopia island (12.3°S, 168.8°E) in the outer Solomon Islands, generated by Tropical Cyclone Zoe in late December 2002. Photo by Geoff Mackley.

is associated with the big drop in atmospheric pressure close to the centre of tropical cyclones. The low pressure causes a temporary but rapid rise in the local sea level. Along affected sections of coastline, this sea-level rise may take an hour to reach its peak as barometric pressure falls to its minimum value, and roughly the same time to go back down again after the cyclone has moved away. The overall effect of strong winds and low pressure is that sea level near the middle of a tropical cyclone rises in a dome of water typically 50 km across, and from 2 to 5 m higher than the predicted height of tide (Table 5.5).

TABLE 5.5. Examples of observed maximum surge heights produced by tropical cyclones in the Coral Sea and the South Pacific.

Place recorded	Tropical cyclone	Year	Surge height above normal sea level (m)
Bathurst Bay, Queensland, Australia	Mahina	1899	13
Mackay, Queensland, Australia	Unnamed	1918	6
Southern Viti Levu island, Fiji	Unnamed	1941	1.8
Funafuti atoll, Tuvalu	Bebe	1971	4
Nayau island, Fiji	Meli	1979	2–3
Tikehau atoll, Tuamotu archipelago	Veena	1983	4
Beqa island, Fiji	Oscar	1983	3–4
Southern Viti Levu island, Fiji	Hina	1985	1
Rarotonga, Cook Islands	Sally	1987	5

The extra rise in sea level produced by storm surges means that coral reefs, which normally afford protection around tropical island shores, sometimes become deeply submerged. This allows the enormous waves to attack exposed locations, leading to a variety of shoreline erosional and constructional changes (Nunn 1994). Percolation of salt water into the soils of coastal plains also kills vegetation and contaminates natural freshwater aquifers, particularly in wells (Fig. 5.18).

FIG. 5.18. Stagnant sea water standing on low-lying areas of the main islet on Pukapuka atoll in the northern Cook Islands. The atoll was flooded several days earlier by Tropical Cyclone Percy on 26 February 2005. Photo courtesy of Douglas Clark.

5.3.2 Cyclone and Coastline Influences

At any particular coastal location, the level and duration of sea flooding by storm surge depends on the following factors, which are discussed below:

1. Depth of the horizontal atmospheric-pressure gradient across the eye of the storm,
2. Speed and radius of maximum winds,
3. Direction and speed of cyclone motion in relation to the land,
4. Timing of tropical cyclone landfall in relation to local tides,
5. Shape of the coastline near the place of cyclone landfall,
6. Bathymetry of the nearshore zone.

In general, the rise in sea level due to falling atmospheric pressure is about one centimetre for every millibar fall in pressure. This is called the *inverse barometer effect*. The fall in atmospheric pressure associated with tropical cyclones often exceeds 50 mb, so the contribution to the total rise in mean sea level due to the lowering of atmospheric pressure can be over half a metre. The relationship between surge height, maximum wind radius and cyclonic fall in atmospheric pressure is illustrated in Fig. 5.19.

A storm surge will clearly have its greatest potential for swamping low-lying deltas and coastal plains if it coincides with the time of high tide. If the time that a tropical cyclone makes landfall, or the time of its nearest proximity to an island, occurs at the same time as a high tide, then the addition of the storm surge and the high tide produces a *storm tide*, which can rise to a much greater elevation than the normal high tide (Figs. 5.20 and 5.21).

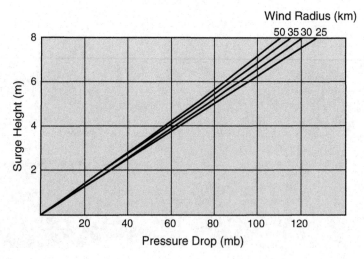

FIG. 5.19. Storm surge as a function of the drop in atmospheric pressure and the radius of maximum winds. This is for a tropical cyclone approaching landfall across a basin of standard bathymetry, perpendicular to a straight coastline at a speed of 25 km h^{-1}. Adapted from Jelesnianski (1972).

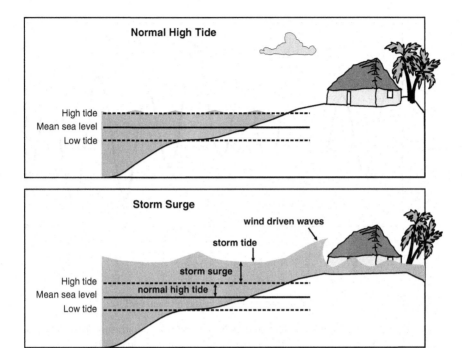

Fig. 5.20. Illustrating the effect of storm surge coinciding with high tide, combining into a storm tide and causing inundation of an island coastline. Adapted from Australian Bureau of Meteorology (2005b).

Faster wind speeds and larger tropical cyclone systems have the potential to inundate a greater length of coastline than smaller or weaker systems. If a cyclone moves rapidly and directly towards a landmass, then there is a good chance that the coast will be flooded. Storm surge can also be dangerous where a tropical cyclone is advancing slowly, but moving parallel or at a low tangent to the shoreline. In the Southern Hemisphere, the full brunt of sea flooding is experienced to the left of the cyclone track in the direction of approach to an island. This is a result of the highest storm surge felt near the cyclone centre, where atmospheric pressure is lowest, in conjunction with ferocious winds just outside the eye pushing waves in an onshore direction. The maximum rise in sea level from surge and wind-driven waves is felt from 20 to 50 km to the left of the storm track as the cyclone passes, as shown in Fig. 5.22.

Long island coastlines with many indentations and a gradually sloping off-shore sea bed are more vulnerable to surge conditions. This is because there is less opportunity for the surge waters to be evacuated along the coast or around the sides of the island (Krishna 1984). As cyclone-generated swells and waves approach land, the influence of the sea bed is to cause friction at the bottom of the water column, especially if the offshore slope is gentle. This interferes with returning water currents at depth, and assists the strong winds at the sea surface to heap water up against the shore. A bay or river

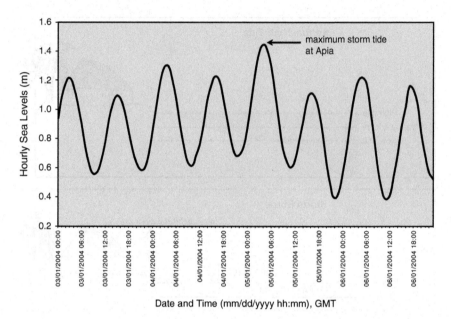

Date and Time (mm/dd/yyyy hh:mm), GMT

FIG. 5.21. Graph of high tide measured at Apia in Samoa, illustrating the storm tide produced by Tropical Cyclone Heta at 4 a.m. GMT on 5 January 2004.

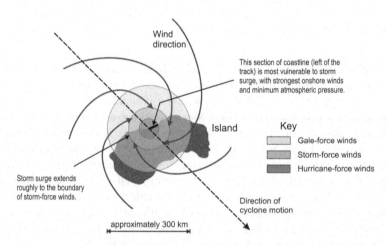

FIG. 5.22. Landfall of an idealised tropical cyclone at an island, indicating the section of coastline that suffers maximum storm surge from the combination of low pressure and fierce onshore winds; note: eye not shown for clarity. Adapted from Krishna (1984).

estuary in the path of the piled-up water produces a funnelling effect. Water is effectively pumped into the semi-enclosed space where it is trapped, so surge heights within inlets can reach levels considerably higher than along an open coastline.

5.3.3 Case Study – Storm Surge Produced by Tropical Cyclone Gavin in Fiji, March 1997

A fine illustration of the effects of storm surge on sea flooding is provided by the example of Vanua Levu island in Fiji during Tropical Cyclone Gavin in early March 1997 (Terry and Raj 1999). TC Gavin, lasting from 3 to 11 March, was the first cyclone to strike the Fiji Islands in the 1997 hot season (Fig. 5.23), and was the severest storm to affect the nation since the early

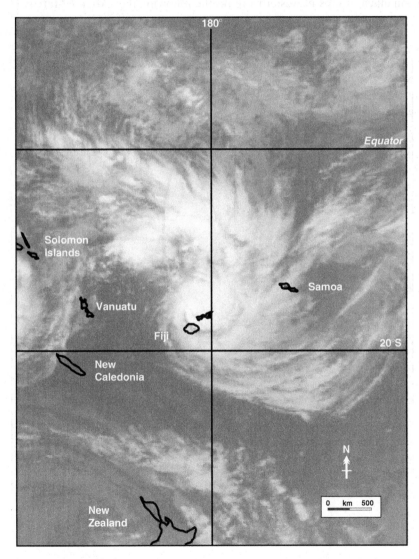

FIG. 5.23. Satellite image of Tropical Cyclone Gavin, 8 a.m. local time on 7 March 1997. Note the well-developed structure of TC Gavin in its mature stage, especially the spiralling arms of cloud. Base image courtesy of the Fiji Meteorological Service.

1990s (Fiji Meteorological Service 1997b). Initially the depression formed north of Fiji waters and west of the atolls of Tuvalu, achieving tropical cyclone status with storm-force winds at a position approximately 10°S 173°E.

By the evening of 5 March, TC Gavin strengthened to hurricane intensity, with winds gusting to 130 knots (240 km h^{-1}). The system approached the large island of Vanua Levu from the north on 6 March, but shortly after midnight it altered course to the southwest (Fig. 5.24). Continuing on this track, the storm traversed directly over the small islands of the Yasawa and Mamanuca groups in western Fiji on the following day. After 7 March, TC

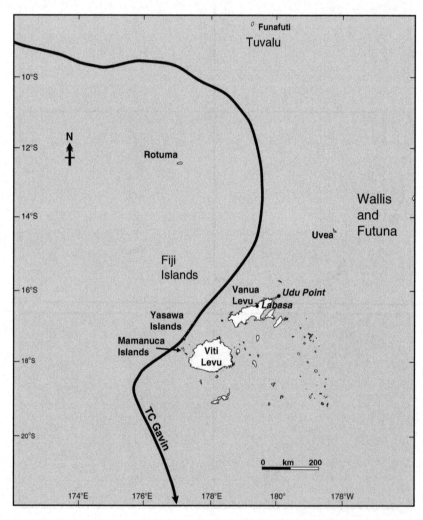

FIG. 5.24. Track of Tropical Cyclone Gavin through Fiji waters, from 4 to 10 March 1997.

Gavin maintained a southerly track away from the Fiji archipelago, but stayed at hurricane strength until well after leaving Fiji waters.

A heavy storm surge from TC Gavin inundated many coastal areas in northern Fiji. On Vanua Levu island, almost the total length of the north coast was affected, with sea walls breached in at least ten places (Fig. 5.25). Labasa town, the largest urban area on Vanua Levu, sits on the banks of two major rivers, the Labasa River and the Qawa River. Surface water levels in the estuary of the Labasa River are measured continuously by the Hydrology Division of the Fiji Public Works Department. In Fig. 5.26 the water levels during TC Gavin are compared with barometric pressure measured at the Udu Point weather station, situated 75 km away on the northeastern peninsular of Vanua Levu.

For the period that TC Gavin was approaching from northern Fiji waters, Labasa was on the right side of the cyclone track, relative to its direction of movement. This means that winds were blowing offshore as the storm advanced, which had the effect of driving water out of the river estuary. This produced an exceptionally low tide at 10 p.m. on 6 March. Once the eye of the storm had passed the mouth of the Labasa River on the following morning, the winds then swung around to an onshore direction. These vigorous onshore winds effectively retarded the outflow of water during the next low tide at 1 p.m. on 7 March, giving a low tide that was barely lower than normal

FIG. 5.25. High seas spilling over the sea wall at Tabucola on the northern coast of Vanua Levu island in Fiji, pushed up by storm surge associated with Tropical Cyclone Gavin on the afternoon of 7 March 1997. Photo courtesy of the Land and Water Management Division, Fiji Ministry of Agriculture.

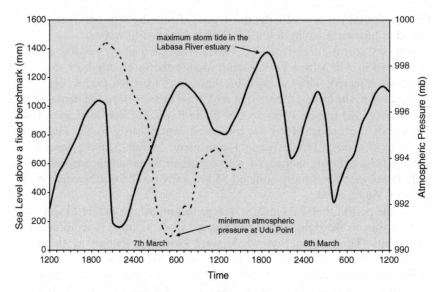

F<small>IG</small>. 5.26. Water level in the Labasa River estuary, measured at Labasa town on the north coast of Vanua Levu island in Fiji, during Tropical Cyclone Gavin in March 1997. Water level is compared with corresponding low barometric pressure at Udu Point on the northeast peninsular of Vanua Levu island, 75 km distant. The maximum storm tide peaked around 7 p.m. local time on 7 March and caused widespread coastal flooding.

high tide. At the time of the next high tide at 7 p.m. in the evening, the storm tide caused comprehensive flooding of Labasa town and the surrounding flat coastal hinterlands.

5.4 Torrential Rainfall

5.4.1 Distribution and Controls

Tropical cyclones normally produce torrential rainfalls, meaning that both the amount of rainfall is large (total in mm) and the rainfall intensity is high (amount per unit time, expressed in mm h^{-1}). Over oceans the rainfall is of convectional type since there are no fronts associated with tropical cyclones. Converging air rises over warm sea and cools adiabatically, leading to the condensation of water droplets, cloud formation and subsequent rainfall. But precipitation near the centre of tropical cyclones is far in excess of the moisture supplied by local evaporation from the ocean surface (Liu *et al.* 1995), and benefits also from the influx of moist air drawn in from the surroundings. The transport of moisture by horizontal winds is known as *moisture advection*. Moisture advection from converging winds results in significant *moisture convergence* near the centre of the system, producing thunderstorm development.

It is a popular misconception that tropical cyclone intensity is a good guide for the expected precipitation. This is not the case. Weaker systems are just as capable of producing huge rainfalls as very intense systems. For any particular island location, the precipitation received during a tropical cyclone depends on several factors. First, rainfall generally tends to decrease in a linear manner away from the eyewall region (Riehl 1954), so the closer the proximity of the storm track to an island, the greater the chance of high rainfall there. Second, precipitation falls in well-defined bands rather than uniformly across the middle to outer regions of the storm. This is due to the horizontal cloud pattern described earlier, organised in spiral-shaped arms of cloud with clearer weather in between. The term *moat* is used to refer to the area of relatively light cloud and rain between the eyewall and the arms of cloud farther out where rain can also be heavy. Within the rain bands are smaller cells of intense precipitation that can only be detected by radar (Fig. 5.27). The spiral bands of cloud and

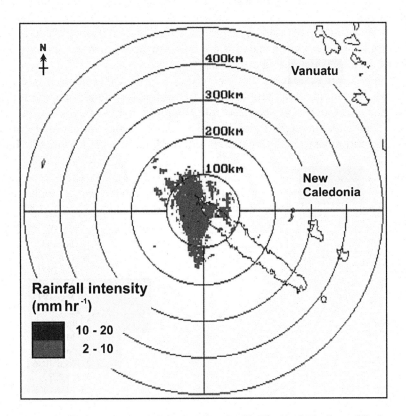

FIG. 5.27. Radar image of rainfall associated with Tropical Cyclone Olinda over Grande Terre island, New Caledonia, on 22 January 1999. The radar defines very small cells of precipitation with intensity 10–20 mm h^{-1} (points of darkest shading) within a larger area of rain falling at 2–10 mm h^{-1} over the northern part of the island. Original base image courtesy of the Australian Bureau of Meteorology.

uneven distribution of rain cells within them are responsible for big localised differences in precipitation receipt across a tropical cyclone.

A third factor is the combination of speed and direction at which a tropical cyclone approaches an island. A fast-moving system approaching an island directly may pass by quite quickly and so deliver only modest rainfalls. In contrast, if a stationary or slow-moving storm lingers near an island for an extended period, then heavy rain will be prolonged and may deliver record-breaking precipitation totals.

A fourth and very important factor is the influence of island elevation and topography, because orographic effects add significantly to the amount of air-mass cooling and moisture condensation. The orographic influence is not felt on low-lying atolls or limestone islands, but if a cyclone travels past a high volcanic island, the steep terrain forces orographic uplift of the spiralling rain bands. This can lead to the production of truly enormous rainfalls, especially at high elevations and on the windward sides of mountains, with respect to the direction of the storm track (example given in Section 5.4.2). Torrential orographic downpours in highland interiors of islands during tropical cyclones can have severe consequences for slope stability and river flooding, which are discussed in Chaps. 8 and 9.

A simple way to compare between successive tropical cyclones is to examine their maximum 1-day rainfall. Climate stations use standard apparatus to record rainfall over 24-h periods from 9 a.m. to 9 a.m. Tables 5.6 and 5.7 show such data for single stations in New Caledonia and Vanuatu. At Liliane climate station near New Caledonia's international airport on Grande Terre

TABLE 5.6. List of maximum daily rainfalls greater than 50 mm (arbitrary threshold), produced by tropical cyclones from 1992 to 2002, at Liliane climate station near Tontouta international airport, southwest Grand Terre island in New Caledonia.

Tropical cyclone	Maximum 1-day rainfall[a] (mm)	Date of measurement
Betsy	77.5	10 Jan 1992
Daman	121.0	16 Feb 1992
Esau	173.5	4 Mar 1992
Fran	110.5	23 Mar 1992
Roger	62.5	14 Mar 1993
Rewa	84.5	5 Jan 1994
Theodore	104.5	27 Feb 1994
Usha	50.5	28 Mar 1994
Zaka	76.5	10 Mar 1996
Beti	374.5	27 Mar 1996
Drena	188.5	7 Jan 1997
Yali	64.0	23 Mar 1998
Zuman	59.5	5 Apr 1998
Dani	65.5	22 Jan 1999
Frank	64.5	16 Feb 1999
Des	129.0	6 Mar 2002

Source: Observatoire de la Ressource en Eau, DAVAR, Noumea, New Caledonia.
[a]Measured to the nearest 0.5 mm.

TABLE 5.7. List of maximum daily rainfalls greater than 50 mm (arbitrary threshold), produced by tropical cyclones from 1981 to 2003, at Pekoa airport outside Luganville town on Santo island, Vanuatu.

Tropical cyclone	Maximum 1-day rainfall (mm)	Date of measurement
Cliff	92.4	15 Feb 1981
Gyan	112.8	23 Dec 1981
Kina	52.7	4 Nov 1982
Beti	75.4	6 Feb 1984
Eric	100.8	15 Jan 1985
Nigel	246.4	18 Jan 1985
Hina	52.8	14 Mar 1985
Keli	122.0	13 Feb 1986
Patsy	112.5	15 Dec 1986
Uma	131.3	5 Feb 1987
Anne	247.6	11 Jan 1988
Bola	262.2	29 Feb 1988
Eseta	87.1	18 Dec 1988
Lili	171.5	8 Apr 1989
Betsy	77.6	9 Jan 1992
Daman	117.5	15 Feb 1992
Esau	219.3	25 Feb 1992
Prema	57.7	29 Mar 1993
Sarah	266.6	24 Jan 1994
Tomas	127.5	25 Mar 1994
Fergus	65.0	26 Dec 1996
Susan	78.8	6 Jan 1998
Yali	98.0	20 Mar 1998
Zuman	223.7	1 Apr 1998
Dani	148.0	19 Jan 1999
Ella	107.7	5 Feb 1999
Iris	115.0	10 Jan 2000
Paula	119.6	27 Feb 2001
Sose	129.6	7 Apr 2001
Gina	115.0	9 Jun 2003

Source: Vanuatu Meteorological Service, Port Vila, Vanuatu.

island, 16 cyclones in 11 years produced 1-day rainfall over 50 mm. Nearly half of these storms delivered daily rainfalls in excess of 100 mm. At Pekoa climate station on the south coast of Santo island in Vanuatu, 30 cyclones in 23 years produced 1-day rainfall over 50 mm; 67% and 20% of storms delivered daily rainfalls greater than 100 and 200 mm, respectively.

5.4.2 Case Study – Rainfall Distribution Across Fiji During Tropical Cyclone Gavin in March 1997

The track and behaviour of Tropical Cyclone Gavin through the Fiji Islands from 4 to 11 March 1997 were described earlier in Section 5.3 on storm surge. Meteorological records from Fiji's 22 synoptic climate stations demonstrate the widely variable pattern of rainfall delivered by TC Gavin over the islands.

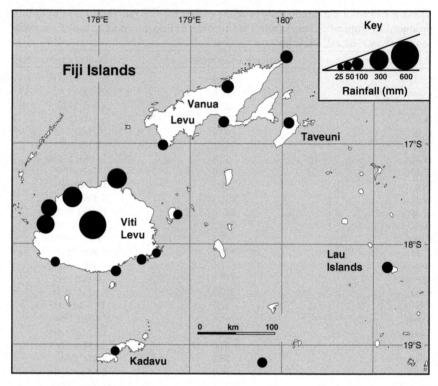

FIG. 5.28. Maximum 1-day rainfall delivered by Tropical Cyclone Gavin across the Fiji Islands in early March 1997.

The information is summarised in Fig. 5.28 that illustrates the distribution of maximum 1-day rainfall. It is seen that huge amounts of moisture were received along the northwest coast of Viti Levu. This is because TC Gavin approached from the north and slowed down as it moved past the island. The interior highlands of Viti Levu experienced the greatest deluge of all, owing to orographic lifting of the storm's rain bands. At an elevation of 760 m in the central mountains, the climate station at the Monasavu hydroelectric dam measured a phenomenal 1-day downpour of 615 mm on 7 March. This value set a new long-term record for 1-day rainfall in Fiji. Associated maximum rainfall intensities reached a drenching 152 mm h^{-1}, calculated over 10-min intervals from traces on automatic rain-gauge charts.

Chapter 6
Future Tropical Cyclone Activity

6.1 Methodologies

If we are interested in understanding something about the sensitivity of island environments to the impacts of tropical cyclones, which is the focus of the following chapters of this book, then we must at some point tackle the thorny issue of what, if any, will be the effects of climate change on the future characteristics and behaviour of tropical cyclones. As might be imagined, in all of the world's major ocean basins, with the South Pacific no exception, questions abound concerning the nature of the tropical-cyclone regime in a projected warmer world. The surfaces of most tropical oceans have warmed up by 0.25–0.5°C during the past few decades (Santer *et al.* 2006) and it is widely believed that the increase in greenhouse gas concentration is the primary cause of the observed rise in global mean sea-surface temperature over the past 50 years (IPCC 2001). In cyclone research, scientists have to date largely concentrated their attention on assessing the likelihood of any climate change and associated ocean warming effects on tropical cyclone numbers and frequencies, storm intensities, and the locations of storm origins. Tropical cyclone durations and precipitation have also featured as topics for research, albeit not so prominently.

Several different methodologies for projecting future changes in tropical cyclone activity have been employed, including empirical, theoretical and physical modelling. Empirical modelling examines the available statistical data on past tropical cyclones, being careful to take into account the bias that exists in historical records prior to the advent of the satellite age in the early 1970s. Any observable trends are then extrapolated into the future. Theoretical modelling, in comparison, is a rather more complex task, since it involves making predictions based on those known laws of physics that are relevant for tropical cyclogenesis, such as the laws of thermodynamics. These laws can be boiled down to sets of mathematical equations. Predictions about the nature of future tropical cyclones are then derived by adjusting the values in these mathematical equations, and then calculating (and interpreting) the results.

Physical modelling is based on simplifying the physical processes that are responsible for the generation of tropical cyclones. Future storm activity can then be simulated in response to varying environmental parameters such as sea-surface temperatures or atmospheric CO_2 content, or combinations of changing parameters. The following sections summarise some of the main concepts and estimations of future tropical cyclone activity related to climate change, based on statistical analysis, modelling and simulation experiments by leading scientists in the field.

6.2 Changes in Frequency

It is probable that (1) tropical ocean basins have warmed significantly due to the long-term increase in greenhouse gas concentrations (Karoly and Wu 2005), and (2) tropical sea-surface temperatures will continue to rise this century at a faster rate than through the last century, if anthropogenic emissions of greenhouse gases follow the projections of the Intergovernmental Panel on Climate Change (IPCC 2001). It is also understood that tropical cyclones form over warm oceans, and that the warmth of the ocean is the primary control driving cyclone intensification. Since these concepts are simply grasped, one can be forgiven for taking what seems to be a logical step forward from this point, by making the assumption that global warming and the associated rise in sea-surface temperatures should lead to increasing numbers of tropical cyclones.

Yet this assumption is a false one, and on a global scale much uncertainty remains concerning the influence of climate change on tropical cyclone frequencies. Indeed, careful analysis of worldwide tropical cyclone data over the past 35 years by Webster *et al.* (2005) shows that no global trends in the temporal pattern of storm numbers have yet emerged, in spite of a background increase in sea-surface temperatures over the same period. Several earlier studies even suggested that the number of storms is significantly reduced with a doubling of the CO_2 concentration in the atmosphere (Bengtsson *et al.* 1996), especially for the Southern Hemisphere. Many scientists argue that the achievement of greater accuracy in forecasting tropical cyclone incidence is limited by relying on statistical methods alone (Basher and Zheng 1995), but the output from modelling approaches has also yielded mixed results (Oouchi *et al.* 2006).

The message to bear in mind is that the initiation and subsequent maturation of incipient tropical cyclones depends on a suite of variables, not just sea-surface temperatures. This is borne out by the observation that from year to year there is actually no established positive relationship in any ocean basin (except the North Atlantic) between cyclone frequency and sea-surface temperature (Raper 1992, WMO 2006b). Walsh and Pittock (1998) remind us that several other variables strongly influence cyclogenesis, including the vertical lapse rate of the atmosphere, the wind shear environment, relative humidity levels in the mid-troposphere, and other conditions necessary to maintain the existence of a low-level centre of cyclonic vorticity. Therefore, changes in tropical cyclone frequency in the future will probably depend on any modifications to regional-scale circulation patterns influencing all of these parameters, not simply warmer oceans alone (Hulme and Viner 1998, Royer *et al*. 1998).

6.3 Changes in Intensity

Sea-surface temperature (SST) is one important variable to which the intensity of tropical cyclones is sensitive (Evans 1993). It therefore comes as no surprise that several models, based on historical records in various ocean basins that have warmed over recent decades, suggest that increasing SSTs will lead to stronger tropical cyclone intensities (Knutson *et al*. 2004). This is backed up by recently published work, hotly debated amongst the global scientific community, in which evidence showing a long-term increase in tropical cyclone intensity has been documented. Emanuel (2005) presented evidence for a substantial gain in the power of cyclones over time, as denoted by maximum winds. Webster *et al*. (2005) analysed a range of tropical cyclone parameters in different oceans over the last 35 years, and reported that a large increase in storm intensity is observed over this period for the southwest Pacific.

The fly in the ointment, unfortunately, is that there are fundamental problems with using historical records of cyclone intensities that cannot be ignored. The historical record is of course a product of real-time operations that were carried out in the past. Consequently, gradual improvements in satellite technology, data density and quality, and even changes in personnel training, have had a continuous impact on the accuracy of the cyclone record over time (WMO 2006b). So, for example, Landsea *et al*. (2006) conclude that the historical record of maximum wind speeds in tropical cyclones is far from perfect because of the evolution of satellite monitoring capabilities and operational procedures. In consequence, many scientists maintain that 'observed' increases in tropical cyclone intensities can be attributed merely to better cyclone observation techniques and improving data reliability over recent decades (Klotzbach 2006).

Yet in spite of these arguments, any shortcomings in historical records do not cast any doubt on the basic idea that global warming should have an impact on the strength of tropical cyclones, and it is generally believed that tropical cyclone intensities will probably show an increasing trend in future, as the atmospheric content of greenhouse gases rises (Henderson-Sellers *et al*. 1998). The uncertainty lies only in the *magnitude* of the change.

Another approach is to evaluate changes in storm *potential intensity* from thermodynamic principles, the validity of which is supported by the fact that calculated potential intensities and observed maximum intensities of tropical cyclones tend to agree rather well (Tonkin *et al*. 2000). It is understood that there is an upper limit to how much a tropical cyclone can intensify, called the *maximum potential intensity* or MPI. This is because as a tropical cyclone intensifies and its winds grow stronger, the energy losses increase relative to energy gain (Holland 1997). Essentially, the energy gained through evaporation of moisture from the ocean increases in a linear manner with wind speed, but this is outpaced by the amount of energy lost to the ocean surface by friction, which is lost in proportion to the *cube* of the wind speed. Based on these and other principles, Emanuel (2004) calculated that a rise of 2°C in tropical sea-surface temperature increases the maximum potential intensity of tropical cyclones by nearly 10%. One point to remember, however, is that individual tropical cyclones rarely achieve their MPI before they encounter unfavourable atmospheric conditions, cooler sea surfaces or are negatively affected by a landmass.

The use of climate models has also been attempted for forecasting tropical cyclone intensities. Early climate models with coarse resolutions (typically 100 km) were not up to the task because they cannot represent the inner structures of a typical tropical cyclone, such as mesoscale convection. But recent years have seen rapid progress in the capabilities of high-resolution (20–100 km) models, which are proving much more useful.

Using high-resolution modelling experiments, Knutson *et al.* (2004) simulated greater intensity hurricanes in the northwest Atlantic in conditions of higher-than-present atmospheric CO_2. Over a 100-year simulated period, warmed by a 1% per year cumulative build-up of atmospheric CO_2, the maximum winds in storms strengthened by between 3 and 10% and the central low pressures decreased by 8 mb. For the Australian region, Walsh and Ryan (2000) also predicted an increase in future cyclone intensity, although this seems to contrast with the experimental results of Oouchi *et al.* (2006) for the South Pacific region as a whole, which suggested an almost 10% reduction in maximum surface wind speeds over the next 10-year period. Clearly there is much room for further investigation to explain these (modelled) regional differences in future tropical cyclone intensities.

Extrapolating storm intensities into warmer-world conditions is also complicated by the influence of other environmental parameters besides atmospheric CO_2 and sea-surface temperatures. Vertical shearing and atmospheric lapse rates, for example, are critical influences and will undoubtedly be affected by global warming as well (Free *et al.* 2004). Even identifying any changes in average cyclone intensity may prove difficult in the short term, because small incremental increases through time will be masked by the normal statistical variations throughout cyclone seasons and from one year to the next. For this reason it may take several decades before noticeable increases in tropical cyclone intensities begin to be manifested during real storms, rather than just in simulated events (Knutson *et al.* 2001).

Nonetheless, even small increases in the strength of tropical cyclones could result in a disproportionate escalation in damages and socio-economic disruption in the South Pacific islands (Bettencourt *et al.* 2002). Most of the population lives at or near the coast in low-lying areas such as river deltas, emerged Holocene marine platforms or atolls. Increasing vulnerability to sea flooding associated with storm surge is therefore a major concern in the context of increasing tropical cyclone intensity, especially when coupled with projected rises in global sea level by the IPCC (2001).

6.4 Changes in Origins

It is a paradox that although the conditions necessary for generating cyclones are actually present in the tropical latitudes for much of the year, the storms themselves are comparatively rare (Emanuel 2004). Added to this is the fact that our success in forecasting the occurrence of individual tropical cyclones remains poor, largely because of our continuing ignorance of the physics underpinning tropical cyclogenesis. This situation presents an obstacle to predicting future changes in the origin of tropical cyclones.

As a rule, the right conditions to spawn tropical cyclones exist in those regions where the atmosphere is slowly ascending on a large scale. Based on this premise, it is difficult to theorise about any expansion of these regions forced by climate change. The simple reason is that a balance needs to be maintained on a global scale, with approximately equivalent amounts of ascending and descending atmosphere. This means that even if the area of sea-surface temperatures within the bounds of the 26°C isotherm were to grow as a consequence of global warming, it is incorrect to assume that the area of tropical cyclogenesis would expand in a similar fashion (Emanuel 2004).

For the South Pacific region it is clear that El Niño conditions have the effect of spreading the distribution of tropical cyclone origins farther east across to the central South Pacific, compared with average years. A number of climate-change models indicate a shift towards a more El Niño-like state in the South Pacific as greenhouse warming continues (Trenberth and Hoar 1997, Timmermann *et al.* 1999, Houghton *et al.* 2001). It therefore makes sense to suggest that if El Niño episodes become either stronger or more common as a result of climate change, then a more easterly spread of tropical cyclones will become the usual rather than the unusual pattern.

In addition, if the 26°C isotherm is pushed south of its current mean position by ocean warming, then tropical cyclones are likely to have longer lifespans and be able to travel farther polewards before they lose energy and decay. This may be assisted by storm tracks following more southerly directions, as a consequence of more southward-steering winds in enhanced greenhouse conditions (Walsh and Katzfey 2000).

6.5 Changes in Precipitation

Torrential precipitation is one of the key products of tropical cyclones and can trigger major landscape responses on Pacific islands. It is therefore crucial that potential changes in the characteristics of tropical cyclone rainfall are investigated. Moisture convergence is an important component of the water vapour budget in tropical cyclones. So in principle, if the water vapour content of the tropical atmosphere increases in response to global warming, then the moisture convergence for a given amount of dynamic convergence in a tropical cyclone will be enhanced, leading to higher rates of precipitation (WMO 2006b).

In spite of this easy concept, investigating changes in tropical cyclone rainfall has not proved a popular theme for research. One of the main reasons is the insufficient resolution of many climate models for predicting storm precipitation. For example, Walsh and Pittock (1998) reviewed the ability of several climate models to provide insight into enhanced greenhouse effects on the climatology of tropical cyclones and accompanying rainfall extremes. Findings did show an overall tendency for more heavy rain, but poor resolution of climate models and their generally crude representation of convective processes meant that little confidence could be placed in any predictions.

More recently, Knutson et al. (2004) carried out simulation experiments of hurricanes in the North Atlantic, under warmer global conditions with elevated levels of atmospheric CO_2. The aggregated results of many simulations indicated an 18% increase in average precipitation rates within 100 km from storm centres, and an enhancement of precipitation totals by 18–32%. Modelling results for the western North Pacific also suggest increasing cyclone rainfalls. Hasegawa and Emori (2005) used a high-resolution general circulation model to simulate precipitation along the tracks of tropical cyclones under present-day and doubled-CO_2 climates. The simulations predicted that a doubling of atmospheric CO_2 boosts cyclone precipitation, attributed to elevated atmospheric moisture contents. It is important to exercise caution with these kinds of results, however, because changes in precipitation are sensitive to the type of climate-change model applied in simulation experiments (Knutson and Tuleya 2004). Also, the extent to which these findings in other ocean basins are applicable to the South Pacific still remains to be determined.

6.6 Outlook

As seen from the preceding sections, some theoretical methods give fairly robust predictions, whereas some modelling techniques have yielded inconclusive findings. A recent report from the World Meteorological Organization neatly summarises our existing knowledge on the theme of global tropical cyclones and climate change (WMO 2006c).[1] The report lists a number of 'consensus statements' by leading scientists, which include the following:

- Though there is evidence both for and against the existence of a detectable anthropogenic signal in the tropical cyclone climate record to date, no firm conclusion can be made on this point.

---

[1] A detailed statement (WMO 2006b) and a summary statement (WMO 2006c) were released by the World Meteorological Organization following their International Workshop on Tropical Cyclones-VI. This meeting was held over 2 weeks in November 2006 in Costa Rica and was attended by invited participants and prominent scientists on both 'sides' of recent debates about climate change and its impacts on tropical cyclone activity.

- No individual tropical cyclone can be directly attributed to climate change.
- Tropical cyclone wind-speed monitoring has changed dramatically over the last few decades, leading to difficulties in determining accurate trends.
- There is an observed multi-decadal variability of tropical cyclones in some regions ... [which] makes detecting any long-term trends in tropical cyclone activity difficult.
- It is likely that some increase in tropical cyclone peak wind-speed and rainfall will occur if the climate continues to warm. Model studies and theory project a 3–5% increase in wind speed per degree Celsius increase of tropical sea surface temperatures.
- There is an inconsistency between the small changes in wind speed projected by theory and modeling versus large changes reported by some observational studies.
- Although recent climate model simulations project a decrease or no change in global tropical cyclone numbers in a warmer climate, there is low confidence in this projection. In addition, it is unknown how tropical cyclone tracks or areas of impact will change in the future.
- Large regional variations exist in methods used to monitor tropical cyclones. Also, most regions have no measurements by instrumented aircraft. These significant limitations will continue to make detection of trends difficult.
- If the projected rise in sea level due to global warming occurs, then the vulnerability to tropical cyclone storm surge flooding would increase.

For the South Pacific region, work published makes it possible to refine certain points in the above list and to add several contributions. Overall, in a greenhouse-enhanced warmer world characterised by stronger or more persistent El Niño episodes, it seems likely that there will be some evolution in tropical cyclone activity in the South Pacific. As a result, some or all of the following effects may be experienced:

1. Changes to the pattern of cyclone origins, with less clustering and more spreading to the east than at present,
2. Little change in total cyclone numbers or frequency, but generally more storminess east of 180° longitude,
3. Increased tropical cyclone intensities, with lower central pressure and greater maximum wind speeds,
4. Enhanced precipitation,
5. Longer cyclone lifespans,
6. Track directions tending more southerly,
7. Extended track lengths and farther poleward travel before cyclone decay.

For those who remain sceptical, a better understanding of the relationship between tropical cyclones and long-term climate change is currently being advanced by the emerging new field of *palaeotempestology* (Liu 2004). Palaeotempestology is the study of ancient 'tempests' over geomorphological timescales, and so far this new science has mainly been concerned with investigating ancient tropical cyclones (rather than other kinds of storms). Much of the recent success of palaeotempestology has been accomplished by examining the sedimentary preservation of big storm events in suitable depositional environments, such as coastal lagoons, lakes and back-beach areas.

In coastal northern Australia, for example, Nott (2004) has been able to identify and date the occurrence of severe South Pacific palaeocyclones from erosional marks in raised gravel beaches and sedimentary ridges comprising coral rubble, sand, shells and pumice. Where sequences of geomorphic features are preserved in a coastal landscape, careful examination of their structure and composition can also allow interpretation of palaeocyclone frequencies and intensities (Nott 2003). Across the disciplines of climatology and geomorphology, many researchers are now anticipating that palaeotempestological studies will soon provide new insights into the nature of extreme storms in the Holocene, and that this will improve our capabilities of assessing contemporary and future changes in tropical-cyclone regimes.

Part II
Impacts of Tropical Cyclones

Introduction

"Far from island environments being merely static backdrops for human activities, they are themselves continuously changing, not only on the scale of recent geological time, but also on scales measured in years and decades".

(Stoddart and Walsh 1992, p. 2)[1]

Infrequent tropical cyclones of considerable magnitude are the chief natural cause of environmental disturbance on oceanic islands in the tropical South Pacific and the most important catalyst for major and often long-lasting adjustments in the physical landscape. The second half of this book explores these cyclone-induced effects throughout the region on a wide variety of island types.

The intensity of tropical cyclones, the proximity of their tracks to island archipelagoes, and other parameters concerned with the characteristics of cyclones and their attendant meteorological conditions have already been described in the first half of this book. These parameters are important influences on the *scale* of any storm-induced change on islands. But a more fundamental control on environmental change must surely be *island type*. This is because whether an insular landmass is a towering volcano, a flat limestone block emerged only a few metres above sea level or a low-lying mound of shingle built on top of a coral reef (Fig. II.1), is the key control governing the *processes* of change and the *nature* of adjustments in the physical environment.

Thus, landscape responses to intense cyclonic rainfall on mountainous volcanic islands, with rugged topography and weathered-clay soils, include landslides on hillsides and sediment deposition in valley bottoms. Limestone islands in contrast have no significant relief (except perhaps coastal cliffs) on which slope failures can occur and no surface drainage channels to

---

[1] This statement is taken from the introduction to their monograph on "Environmental Extremes as Factors in the Island Ecosystem", which was published in 1992 by the Smithsonian Institute in the US as Atoll Research Bulletin no. 366.

FIG. II.1. Examples of three principal types of islands that are found within the South Pacific – volcanic, limestone and low coral islands. Top: Erromango volcanic island, Vanuatu (courtesy of Shane Cronin); Centre: Nauru, a raised limestone island (courtesy of Randy Thaman); Bottom: Elongated low coral islands (motu) on Funafuti atoll in Tuvalu (courtesy of Randy Thaman).

accommodate terrestrial flooding. In consequence, although the natural vegetation may be severely damaged, tropical cyclones may have less geomorphic impact. Low coral islands, such as those that occur on atolls, have the most vulnerable physical environments of all. They are little more than unconsolidated heaps of coralline sands and gravels resting on reef foundations, and are especially prone to overtopping by storm surge and cyclone-driven waves. Associated sediment movement can produce remarkable changes – sometimes the complete obliteration of an island altogether, or on other occasions the creation of entirely new land.

Island sensitivity to physical change is strongly conditioned by the degree to which the existing landscape is in dynamic equilibrium with the frequency of tropical cyclones. Walsh (1977) referred to this concept as 'landscape equilibrium with the cyclone environment'. The idea is a simple one: a severe storm is more likely to produce catastrophic change on an island where one has not struck for a long time, since many of the geomorphic features are ill-equipped to withstand the impact. In contrast, on islands where the effects of tropical cyclones have been felt more often, the contribution of an individual storm event to landscape change may be less significant. This is because the evolution of the island's physical environment is in some degree of balance with regular climatic perturbations.

In the following four chapters examples are drawn from a range of locations across the South Pacific (and sometimes neighbouring parts of the North Pacific too were useful), to illustrate the astonishing diversity of coastal, terrestrial and hydrological impacts of tropical cyclones, and the primary factors that determine the magnitude of the physical responses.

Chapter 7
Coastal Geomorphology

7.1 Coral Reefs

7.1.1 Reef Characteristics

The clear and warm waters of the tropical South Pacific support the abundant growth of both hard (scleractinian) and soft corals. Scleractinian species are corals that secrete hard skeletons of calcium carbonate from seawater. Diverse types of *hermatypic species* are the colonial corals, sometimes called frame-builders, which are the ones responsible for the growth of fringing, patch and barrier reefs around volcanic and limestone islands, and the growth of atoll reefs overlying subsiding volcanic foundations. Other organisms that make up reef ecosystems are important sand producers. The biogenic sand contributes to reef, lagoon and beach sediments. These organisms include foremost foraminifera, calcareous and coralline algae, as well as shelled molluscs, echinoderms, calcareous worms and bacteria. Reef-dwelling parrot fish and other coral-eroding fish species are also prolific living producers of coral sand. As they feed on coral polyps, parrot fish excrete fine carbonate sand that falls into interstices in the reef structure and washes onto the reef flat.

Coral reefs are remarkable features of coastal geomorphology because they are built of living coral framework. But any extreme fluctuations in the marine environment can lead to coral mortality, with significant consequences for geomorphic instability of the reefs and nearby island coasts. The characteristics of fringing, patch, barrier and atoll types of reefs and their geomorphic zones are given thorough coverage by Guilcher (1988) and Nunn (1994), so only an outline of the main features of reef morphology is presented here, as a context for illustrating cyclone-induced change.

In brief, the ocean sides of reefs facing into the prevailing winds and swells have steep seaward slopes, dropping off to deep water at inclinations generally between 30° and 45°. The submarine slopes of leeward reefs are often steeper than windward reefs due to higher rates of talus accumulation on the latter. Living corals cover only the top 20 m or so of the reef, where there is strong enough sunlight to support their symbiotic marine algae, called *zooxanthellae*. The zooxanthellae provide the corals with both their bright colours and much of their food supply through photosynthesis (Wells and Hanna 1992).

Windward reef edges have a characteristic spur-and-groove morphology. This morphology is a feature of reef growth rather than an erosional form. The spurs dissipate wave energy and the grooves provide conduits for transferring water and sediments. The size and spacing of the grooves and spurs is therefore believed to reflect the average energy of the breaking waves. The reef edge may have an encrusting surface of pink *Porolithon* and *Lithothamnion* coralline algae, which thrive in the highly oxygenated conditions of constant surf. Isolated limestone blocks of reef rock often stand upon the edge of the reef surface, broken off the living reef front and emplaced there during violent storms.

Extending back from the reef edge is a flat rock pavement with a thin, discontinuous veneer of coral fragments and foraminiferal sands. Colonies of *Halimeda* spp. calcareous algae and *Porites* spp. microatolls are common across the reef flat, whereas patch corals inhabit portions of lagoons, pools and shallow depressions. Striations on the pavement are normally aligned with the reef-edge grooves, indicating the directions of water and sediment flow across the reef flat (Nunn 1994). The lagoon edge of reefs has an irregular slope of corals and sediment cover, descending gently to the sandy bottom of the lagoon.

7.1.2 Reef Damage and Recovery

Tropical cyclones have a variety of geomorphic impacts on reefs. Delicate corals may be devastated, not just immediately but also in the weeks afterwards from secondary effects. Living corals on the seaward edges of reefs face the full brunt of large waves generated out at sea (compared to more sheltered lagoon-side reefs which are exposed to smaller waves with short fetches, agitated by winds over lagoons), and may sustain enormous damage by mechanical destruction. Many coral skeletons are simply smashed into pieces *in situ* by direct wave action (Cooper 1966). It is not unusual to see large coral heads torn from the reef and thrown into shallow water, or entire fields of branching corals broken to bits.

The spur-and-groove morphology of the reef edge can act as a baffle to dissipate some of the wave energy and therefore limit structural reef damage more effectively than in areas of smoother reef topography (Scoffin 1993), but in the strongest cyclones the spur-and-groove morphology may be obliterated, exposing a truncated reef surface. Curved fractures may develop along the reef margin, probably resulting from the heavy concussions inflicted along the whole reef by huge breaking waves. On Rangiroa atoll (15.0°S, 147.7°W) in the Tuamotu archipelago of French Polynesia, reef-edge fractures with radii of curvature measuring 100–200 m were observed

(Bourrouilh-Le Jan and Talandier 1985). Superficial geomorphic changes on reef-flat areas include abrasion, scour hollows and erosion channels.

The amount of cyclone damage to corals is species-selective because the vulnerability of colonies to wave impact is a function of their shape, strength of their skeletons, anchoring tenacity, orientation and mutual buttressing (Done 1992, Scoffin 1993). The branching corals, of which *Acropora* species are most common on South Pacific reefs, are easily fragmented and may disappear entirely from the hardest-hit sections of the outer reef edge. Foliaceous, platy and tabular corals such as *Echinopora* and *Montipora* are likewise not at all resilient, and are snapped off and overturned (Fig. 7.1, Terry 2004b). Delicate species may be provided some shelter in the lee of massive colonies, but are demolished if their neighbour topples over. Only the massive forms such as *Porites* spp. have a better chance of survival. Solitary mushroom corals like *Cycloseria, Diaseria* and *Fungia* are easily rolled along and killed. Other corals are uprooted and subsequently carried into deep water, leading to their eventual demise.

Even corals managing to retain a hold on solid foundations may be crushed by lumps of reef debris landing directly on top of them, or are abraded by swells dragging rubble over their surface. Elsewhere, coral heads are buried by storm sediments and grooves on the reef flat may be plugged by coral debris. The availability of boulders, gravel and sand is therefore important in governing the extent of damage by bombardment, abrasion or

Fig. 7.1. Tabular coral heads overturned by powerful waves during Tropical Cyclone Paula on 4 March 2001, south west Viti Levu island, Fiji.

burial (Scoffin 1993), and is controlled by the age of the coral stand and the time elapsed since the last episode of major disturbance.

At depths below the direct impact of waves, delicate corals may be destroyed by the process of *underwater avalanching*, as described on Tikehau atoll (15.0°S, 148.2°W) in the Tuamotus by Harmelin-Vivien and Laboute (1986). Avalanching occurs in locations where the reef flat is narrow, the fore-reef area lies within 15 m of water and the reef slope is very steep (>45°). Tikehau was affected by a sequence of six cyclones that affected French Polynesia during the 1982–1993 hot season. This was an unusually high level of cyclone activity for this part of the central South Pacific. On the fore-reef area, massive species like *Porites* and *Montipora* were uprooted. Since they were growing on high-angle substrates, heavy chunks of broken remains rolled downwards, sweeping away 50–100% of fragile colonies in deeper water (Intes and Caillart 1994). The destruction increases with depth as more rubble is produced, giving the avalanche effect. The avalanche phenomenon does not occur on gentle slopes (<20°) or where a fore-reef terrace extends forward from the reef front.

Reefs around high islands suffer from additional problems when tropical cyclones strike. The muddy discharge from rivers that are heavily swollen by the intense rainfall on land causes a rapid drop in the salinity of seawater. This combines with the smothering effect of the silt-laden waters. These impacts are worst felt on fringing reefs near river estuaries because here the seawater is most contaminated by the muddy runoff, but barrier reefs farther out to sea can also suffer if strong currents direct the turbid freshwater plumes offshore. For those corals not killed immediately, the stress induced by such acute changes in environmental conditions renders them uncommonly vulnerable to post-storm disease or infestations of pests, such as urchins or crown-of-thorns starfish. Physiological stresses may take a further heavy toll by causing corals to expel their symbiotic algae, becoming bleached and dying later through starvation. For corals that do survive, coral bleaching after the death of the zooxanthallae probably slows the rate of calcification and reef growth, and reduces the protection afforded by reefs to islands.

The overall short-term devastation of coral reefs by tropical cyclones can be very dramatic (Fig. 7.2). The breakdown of coral skeletons interrupts active reef growth for an undetermined period of time, but also provides debris that accumulates as carpets of sediment or ridges on the reef pavement, later to become incorporated into reef-top rubble islets. These features are dealt with under Sections 7.2 and 7.4 on shoreline construction and coral islands. Some debris may also wash completely over the reef flat, especially if the reef is narrow, and accumulate in back-reef lagoons as wedge-shaped lobes. On submarine terraces or at the bottom of fore-reef slopes, cones of coral talus build up. Talus cones may be veneered by fine sediments, which take longer to settle out once seas become calm again in post-cyclone fair weather. Fine detritus will also drop out of suspension and fall into interstices and larger cavities in the reef framework, forming a graded layer of

Fig. 7.2. Complete devastation of corals on the outer edge of a fringing reef on the east coast of Ovalau island in central Fiji, after the impact of Tropical Cyclone Kina in January 1992. The large *Porites* colony in the upper centre of both images is the point of reference to provide orientation. Photos by Ed Lovell.

sediment that is coarse below but fines upwards. In deep lagoons, carbonaceous mud drapes are laid down.

Determining the long-term impacts of cyclones on coral reefs is more problematic, since this requires repeated surveys over time. The large and growing literature on this subject highlights that regrowth responses vary a great deal, according to the location and types of cyclone damages sustained (Connell 1978, Stoddart 1985, Lovell 2006 pers. commun.). To simplify the picture, there are two principal recovery scenarios. One idea is that over time, repeated damage from cyclones leads to changes in the overall make-up of the reef ecosystem, with a reduction in coral biodiversity. If fast-growing but delicate species are selectively removed, and replaced by globular forms of corals that grow more slowly, then the pace of reef development will experience a long-term decline.

There is another possibility. If much of the fragmented reef debris falls immediately down the fore-reef slope and rests there as a submarine apron of talus, this then offers a suitable foundation where new coral colonies can be established and thrive during the post-storm decades of reef recovery. The recruitment of new corals depends on the existence of a species reservoir of living corals nearby. Species reservoirs can be provided by sheltered reefs on the leeward sides of an island, patch reefs in lagoons that survived the cyclone relatively unscathed or from corals on the affected reef that survived at depths below the action of the storm waves. In this way, the seaward extension of the reef is actually assisted by infrequent high-magnitude storm events as a part of the natural process of reef growth – accumulation of talus foundations – renewed reef growth once again. These opposing scenarios of long-term tropical cyclone effects on cycles of reef die-back and renewal pose interesting questions for future research.

7.2 Coastal Erosion

7.2.1 Erosional Features

Scour and removal of unconsolidated beach materials are commonly observed around the coasts of islands following tropical cyclones. These processes are independent of whether the beaches form shorelines on volcanic, limestone or coral islands. Shorelines in soft sediments may also retreat landwards, with low vertical steps cut as temporary features into the beach. The loss of sandy beaches is a major geomorphic change for a coastline, and one which is considered highly undesirable for South Pacific nations that rely on beaches to attract international holidaymakers and sustain tourism-dependent economies (Fig. 7.3).

Several geomorphic consequences of beach removal are apparent, one of which is the exhumation of underlying *beachrock* (Fig. 7.4). Beachrock is a type of calcarenite, meaning it is sandstone made of calcareous fragments. It lithifies beneath the surface of the beach by chemical cementation of the beach materials. Beachrock surfaces are scalloped and pitted by solution, and their exposure is evidence that part of the overlying beach profile has been eroded. The beachrock itself may have slabs plucked from it during cyclones, and trenches may be excavated by scouring in the lee of beachrock outcrops.

In the Kingdom of Tonga for instance, Tropical Cyclone Isaac in March 1982 caused severe beach erosion on many islands (Woodroffe 1983). On Lifuka island (19.8°S, 174.4°W) in the central Ha'apai islands it was estimated

FIG. 7.3. Sandbags installed as a (belated) temporary-protection measure for beachside coconut palms at the Shangri-La Fijian Resort in southern Fiji. The beach was eroded several days earlier during Tropical Cyclone Paula on 4 March 2001.

that 28 m^3 of sand was removed per metre along the beachfront. Elsewhere in the Ha'apai group, freshly exhumed white limestone at the basal sections of low cliffs indicated many places where significant amounts of beach sand had been removed by the storm. In southern Tonga, shoreline positions receded around many of the small sand cays located on the barrier reef north of the Tongatapu mainland, such as 2 m of shore retreat measured on Manima islet. Overall, TC Isaac resulted in the widespread transformation of sandy beaches to beachrock, similar to the examples of Tufaka and Malinoa islets shown in Fig. 7.5.

Another consequence of sediment removal is degradation of coastal features such as spits and bars. On Yanuca island in Fiji, lying adjacent to the southwest coast of the Viti Levu mainland, a narrow sand spit used to be attached to the northern end of the island and extended for several hundred metres. Beach erosion by strong waves during TC Sina in late November 1990 breached the spit, leading to its permanent separation from Yanuca (Fig. 7.6). The altered drainage pattern in the lagoon behind the spit, resulting from the new configuration of the coastline, now impedes tidal-current flow. This effectively prevents entrainment of sand from the lagoon floor and has been one of several factors leading to recent problems of eutrophication and aggradation in the lagoon (Terry *et al.* 2006a).

FIG. 7.4. Exposed beachrock has become a common sight around the coastlines of many Pacific islands, indicative of the loss of overlying beach sand, often as a result of shoreline erosion during heavy storms. This photo was taken on Majuro atoll in the Marshall Islands in May 2004.

Cliffed coastlines around emerged limestone islands and tall volcanic islands may be relatively more resistant than beaches to the erosive power of stormy seas during tropical cyclones, but they are nonetheless still subject to undercutting and collapse. For example, as TC Isaac swept through Tonga in March 1982, several low limestone islands with cliffs standing 2–3 m high had blocks of limestone up to 2 m in diameter detached from above modern wave-cut notches and dropped onto the reef flat (Woodroffe 1983). Figure 7.7 illustrates the severe coastal erosion near Apia on Upolu island in Samoa caused by wave attack during TC Ofa in early February 1990. On the elevated limestone island of Niue, immense waves driven ashore by TC Heta on 6 January 2004 eroded bedrock cliffs, entrained coarse sediments

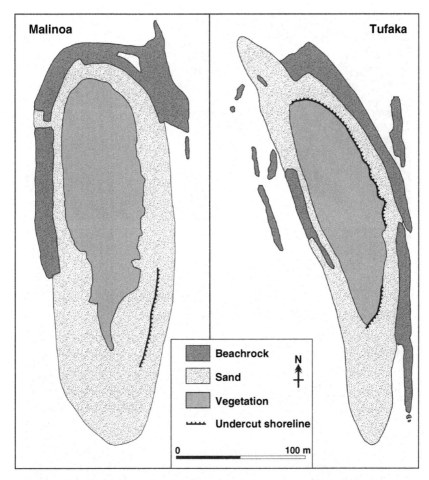

FIG. 7.5. Exposures of beachrock around the sand cays of Tufaka and Malinoa, north of Tongatapu island in the Kingdom of Tonga, after the passage of Tropical Cyclone Isaac on 3 March 1982.

from the narrow fringing reef and carried these onto the top of the coastal marine terrace at a height of 23 m above the sea level.

7.2.2 Case Study – Coastal Erosion on Niue Island During Tropical Cyclone Heta in January 2004

Tropical Cyclone Heta was the first tropical cyclone to form in the RSMC-Nadi area of responsibility during the 2003–2004 tropical cyclone season (Fiji Meteorological Service 2004). The system was first identified as a tropical depression north of Fiji on 28 December 2003, as it moved northeastwards to reach a position just west of Atafu, the northernmost of the Tokelau atolls.

FIG. 7.6. Changes in the geomorphology of a sand spit, formerly connected to the low limestone island of Yanuca in southern Fiji. The spit was breached at its southern end and disconnected from Yanuca island during Tropical Cyclone Sina in late November 1990. In consequence, the narrow lagoon behind the spit has been infilling with sediment since that time.

By this time it had reached tropical cyclone status, and was named at around 3 a.m. on 2 January 2004 (Fig. 7.8). Once mature, the cyclone rapidly strengthened while turning slowly southwards, attaining storm intensity around noon on 2 January and hurricane intensity around 6 a.m. on 3 January. TC Heta was a very intense system with maximum sustained winds estimated at 115 knots (213 km h^{-1}) and momentary gusts up to 160 knots (296 km h^{-1}). Peak intensity was reached about midnight on 5 January while

FIG. 7.7. Coastal erosion east of Apia on Upolu island in Samoa, caused by storm waves accompanying Tropical Cyclone Ofa in February 1990. Source: Rearic (1990).

the cyclone centre was passing 115 km to the west of Savai'i island in Samoa. This strength was maintained over the next 24 h as the cyclone veered onto a southeastward track, its speed accelerating to 15 knots (28 km h^{-1}), and later to 20 knots (37 km h^{-1}). This path took the cyclone centre to within 80 km of Niuatoputapu island in northern Tonga around noon on 5 January and 50 km off the west coast of Niue around 3 a.m. on 6 January.

The hardest-hit country was Niue. Niue is a single isolated island (19.1°S, 169.6°W) located in between Tonga and the southern Cook Islands. TC Heta's hurricane-force winds, enormous seas and associated coastal flooding, caused devastation that was the worst in living memory. Monstrous waves were the chief cause of the destruction. Much of the infrastructure on the island was destroyed and buildings and houses were either demolished or badly damaged, including Niue's hospital complex and parliament. Even houses built atop 30 m high cliffs, and others up to 100 m inland, were flattened. Roads were closed, telecommunications and electricity were cut, and crops were spoiled by the furious winds and torrential rainfall. Alofi, the capital, bore the brunt of the storm, with half of the commercial area wiped out. A woman and her 19-month old child were killed by a giant wave smashing through the house in which they were sheltering. Several other people were injured and many were left homeless. A national disaster was declared by the government and media reports put an estimate on the damage at NZ$50 million.[2]

---

[2] Niue is an independent nation but uses New Zealand currency.

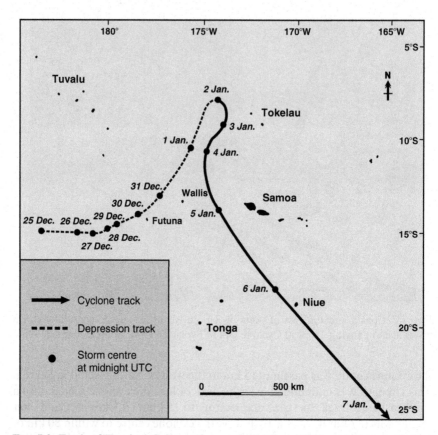

Fɪɢ. 7.8. Track of Tropical Cyclone Heta near Tokelau, Wallis and Futuna, Samoa and Niue in early January 2004.

Niue island is a large raised coral atoll with a land area of 259 km^2 (Murray and Terry 2004). The cliffline comprises hard coral limestones, but the fringing reefs are narrow and afford little protection from rough seas. Cut horizontally into the coastal bedrock is a series of marine terraces at different levels, giving a staircase effect. The terraces indicate the interaction between changing sea levels and tectonic uplift of the island during the Quaternary period. The broadest marine terrace at 23 m elevation is called the Alofi Terrace and encircles the perimeter of the island. Gouged vertically into Niue's coastline are a series of steep-walled chasms and smaller 'sea tracks'. These chasms are straight, sheer-sided, are up to 25 m deep and extend up to 500 m long. They are interpreted as fault-guided solution channels along fault zones that run sub-parallel to the coast (Schofield 1959).

On 6 January 2004, TC Heta's tremendous waves pounded the north and west coasts of the island. Cliffs collapsed and retreated in several places, such as at Makefu on the north coast (Fig. 7.9). The biggest waves were able to dredge coarse debris from the modern-reef surface and overtop the 23 m

Fɪɢ. 7.9. Collapse of a section of limestone cliffs near Makefu on the north coast of Niue island, caused by wave attack during Tropical Cyclone Heta on 6 January 2004. Photo by Michael Bonte.

Fɪɢ. 7.10. Large coral boulder on Niue that was thrown up by gigantic waves during Tropical Cyclone Heta on 6 January 2004. This site is on top of the 23 m elevation marine terrace on the north west coast. Photo by Douglas Clark.

Fɪɢ. 7.11. Hikutavake 'sea track' on the west coast of Niue, swept clean and abraded by waves during Tropical Cyclone Heta on 6 January 2004. Photo by David Talagi.

elevation marine terrace. Large quantities of coral rubble were thrown up onto the Alofi Terrace and deposited there, including many angular boulders weighing several tons (Fig. 7.10). In addition, the coastal chasms and sea tracks funnelled the force of cyclone-generated swells on land. The floors of the chasms were swept clean of sediments and heavily abraded as a result (Fig. 7.11). The funnelling effect also amplified the erosive power of waves, such that coastal forest and buildings up to 200 m inland were torn down (Fig. 7.12).

7.3 Coastal Deposition

7.3.1 Reef-edge Megablocks

The erosional damage by powerful cyclone-generated waves and swells on the seaward margins of coral reefs produces huge quantities of sedimentary debris. The dimensions of some of this debris can be truly spectacular, in the form of massive blocks of limestone ripped from the reef front and then hurled onto the perimeter of the reef platform. On Nukutipipi atoll (20.7°S, 143.2°W) in the Tuamotu archipelago, several giant boulders were lodged onto the reef flat by one of the four tropical cyclones that affected French Polynesia in the 1982–1983 cyclone season (probably TC Orama or TC Veena in early 1983). The chunks of limestone were so big, up to 10 m long, 4 m

Fig. 7.12. Satellite image of the Niue coastline at the capital Alofi on the west coast. Immense waves generated by Tropical Cyclone Heta on 6 January 2004 overtopped the coastal cliffs, and scoured a zone to a distance of up to 200 m inland on the 23 m elevation marine terrace, flattening trees and buildings, including the hospital complex and the Niue Hotel. Base image courtesy of NOAA.

high and 30 m³ in size, that Salvat and Salvat (1992) called them *megablocks*. Of course not all boulders thrown up by cyclonic wave action are necessarily so large, but the occurrence of upstanding blocks perched along reef crests is nonetheless quite common. Through time the blocks get cemented onto the reef surface. In many places the blocks are isolated (Fig. 7.13), but where there are many of them they exhibit sorting, with the largest ones resting close to the reef crest, progressively decreasing in size and numbers farther back towards the lagoon.

7.3.2 Gravel Sheets and Ramparts

Abundant amounts of coral-derived sediment in gravel and sand-size fractions are transported by storm waves onto the surface of reefs. The length of time after the last cyclone and before a subsequent event influences the degree of stabilisation of this material, and hence the chances of its preservation (Scoffin 1993). The coralline sediments are not distributed in random fashion, but normally show distinctive patterns of accumulation. Often the debris is built up in new constructional forms in the coastal landscape. For example, Jaluit atoll (6.0°N, 169.6°E) in the southern Marshall Islands was hit by Typhoon Ophelia on 7 January 1958 (Fig. 7.14) A post-storm survey of several islets on Jaluit by McKee (1959) revealed that typhoon deposits had accumulated as four distinct features: (1) coarse gravel ridges on the reef flat, (2) new or augmented beach ridges on the ocean sides of islets, (3) gravel

FIG. 7.13. Reef-limestone megablock resting on the reef of Suwarrow atoll in the northern Cook Islands. Unconfirmed accounts suggest that the block was lodged on top of its reef platform by a cyclone in 1942. Photo courtesy of Colin Woodroffe.

FIG. 7.14. Fresh coral rubble swept by Typhoon Ophelia onto islets of Jaluit atoll in the southern Marshall Islands on 7 January 1958. Source unknown.

sheets up to a metre thick and (4) new beaches on the lagoon sides of islets (Fig. 7.15) .

Gravel sheets are blankets of deposits laid over the top of islets and reef flats, sometimes several tens of centimetres in thickness (Scoffin 1993). They have

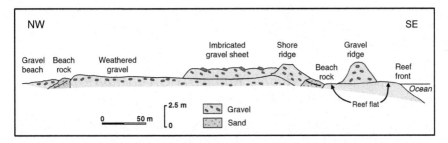

FIG. 7.15. The variety of depositional features (gravel ridge, shore ridge and imbricated gravel sheet) formed on Jabor islet on Jaluit atoll in the Marshall Islands, by Typhoon Ophelia on 7 January 1958. Redrawn from McKee (1959).

mixed compositions of sand and pebbles and are not as coarse as storm ridges. Their front edge may be marked by a vertical face up to half a metre in height. Within gravel sheets, pebbles and cobbles are commonly imbricated, the dip of the clasts showing the direction from which the storm waves advanced.

One of the most significant coastal changes associated with exceptionally violent cyclones is the creation of large *rubble banks* on the reef platform, composed of coral-derived detritus. These features have been variously termed rubble ramparts, gravel ridges, gravel banks, storm embankments or similar combinations of these names. Rubble banks are a major addition to the shoreline geomorphology, but are also very important over the longer term for nourishing beaches and low islands with sediments. Pre-existing ramparts created by earlier storms may be transformed by later cyclones into blanket-lag deposits of coarse particles spread evenly over the reef flat. Rampart geomorphology is discussed here and island nourishment is covered in Section 7.4 on low coral islands.

Two superb examples of rubble bank formation in the Pacific are as follows: A great storm rampart of boulders was created by Typhoon Ophelia in 1958 on the seaward reef flat of Jabor islet on Jaluit atoll in the southern Marshall Islands. The crest of the boulder rampart rose up 2.5 m above the reef flat, about midway between the reef crest and the shore of the main islands (Newell and Bloom 1970). Sediments which were added to the bulk of several islets comprised pebble, cobble and boulder sizes,[3] but with little fine matrix. On Ontong Java atoll (5.3°S, 159.5°E) in the northern Solomon Islands, large and extensive rubble banks were constructed by Tropical Cyclone Annie in November 1967. The ferocity of this storm was so unprecedented in the history of the Ontong Java islanders, that to describe it they introduced a new word 'sakaloni' in their language. An almost continuous

---

[3] Clast sizes in coarse sediments are as follows – pebbles: 4–64 mm; cobbles: 64–256 mm; boulders: >256 mm.

20 m wide rampart between 1 and 3 m high was built between Ngikolo and Lopaha, which are places lying 35 km apart on the southern side of the atoll (Bayliss-Smith 1988).

7.3.3 Case Study – Rubble Ramparts Created on Upolu Island by Tropical Cyclone Ofa in February 1990

Tropical Cyclone Ofa developed over Tuvalu on 27 January 1990 as a depression within the South Pacific Convergence Zone, then rapidly intensified to hurricane intensity as it moved south–southeast. The track was shown in Fig. 5.10. Samoa was severely affected from 2 to 4 February. Very destructive sustained winds greater than 118 km h^{-1} lasted for almost 24 h (Fiji Meteorological Service 1990). The violent winds, rain deluge and storm surge caused widespread devastation across Samoa, the worst in over a century. Enormous waves and high storm tide inundated low-lying coasts as well as many inland areas along river estuaries. The northern coasts of Upolu and Savai'i islands were worst affected.

Along the northern coastline of Upolu island, the force of gigantic waves smashed corals and dredged up huge quantities of debris from the reef. Extensive rubble banks were constructed from this material along the perimeters of fringing and barrier reefs, and storm beaches were deposited along upper shorelines behind narrow fringing reefs (Zann 1991; Table 7.1). The dimensions of the rubble banks were 2–3 m in height and 5–30 m in width, built 10–20 m inshore from the edge of the reef (Rearic 1990; Fig. 7.16). Ramparts were frequently divided by gaps 50–100 m wide at the site of reef passages, but some sections extended continuously to great lengths, up to a maximum length of 1.5 km.

TABLE 7.1. Survey of rubble ramparts and storm beaches created around Upolu island in Samoa, during Tropical Cyclone Ofa on 3 February 1990.

Coastline	Reef front (km)	Number of emergent banks	Length of ramparts (km)		
			Emergent banks[a]	Tidal banks[b]	Storm beaches
West	52	4	1.8	2–3	1
Northwest	65	57	25.5	17	0
Northeast	41	20	12.0	7[c]	4
East	22[c]	5	1.7	2[c]	–
Southeast	41	0	0	–	–
Southwest	50	0	0	2–4	–
Totals	271	86	41	32+	5+

From Zann (1991).
[a]Permanently emerged above high tide.
[b]Submerged at high tide.
[c]Conservative estimate only, as either not visible in aerial photographs or not investigated on the ground.
– Not surveyed.

FIG. 7.16. Rubble bank of coral detritus formed at the reef edge near Apia on Upolu island in Samoa, by storm waves accompanying Tropical Cyclone Ofa on 3 February 1990. Source: Rearic (1990).

Excavations and detailed surveys of the rubble banks around Upolu showed that these features were emergent above high tide level. Seaward flanks of the ramparts had steep angles of repose from 45°–60°. Shore-facing slopes had inclinations approximating 20° and often graded into wide, pre-existing sheets of eroded *Acropora* gravels. Sedimentary composition was largely graded coral shingle with only a minor component of sand, overlying a coarser intertidal bench of boulders (Fig. 7.17).

7.4 Changes on Coral Islands

7.4.1 Coral Island Types

Upon the surface of barrier and atoll reefs form low-lying coral islands known as *cays* and *motu*. These comprise accumulated coralline sand and shingle, often anchored onto an underlying core of cemented coral rubble or reef rock that is slightly emerged above sea level. Unconsolidated low coral islands are ephemeral on geomorphic timespans (Nunn 1994) and are very sensitive to tropical cyclone events of high magnitude, although the geomorphic effects can be constructive as well as destructive.

It is helpful to try to distinguish between cays and motu, although the distinction is not always a sharp one. Cays tend to be formed on patches of reefs

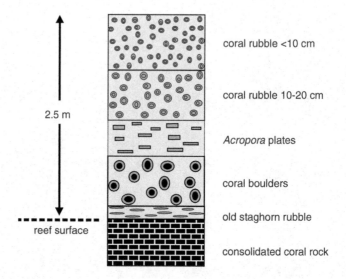

2.5 m

coral rubble <10 cm

coral rubble 10-20 cm

Acropora plates

coral boulders

old staghorn rubble

consolidated coral rock

reef surface

FIG. 7.17. Internal facies structure through the thickest section of the rubble bank constructed on the fringing reef off Mulinu'u Point, Apia, on Upolu island in Samoa, by Tropical Cyclone Ofa on 3 February 1990. Based on original data in Zann (1991).

or at the edges of reef passages by the refraction of waves, and are generally small islets, of the order of a couple of hectares or so in size (sometimes smaller), made of sand and gravel. The Polynesian word *motu* refers instead to the elongated narrow islands found on atoll reefs. Some motu[4] are actually just small islets with similar dimensions to cays, although others may extend several kilometres along the top of the atoll. On South Pacific atolls the motu are frequently divided from one another by surge channels called *hoa*. Some atolls have many motu, almost entirely encircling an atoll lagoon, whereas other atolls have broad gaps between more widely spaced islands, even if the reef itself is continuous.

There are many variations, but since motu are the only land rising above sea level on atolls, the geomorphology of such islands can therefore be considered as entirely 'coastal' in nature. Motu are formed of coarser materials than sand cays, predominantly a mixture of sands, gravels and cobbles. There is a distinct zonation in sediments across a motu from the ocean edge to the atoll-lagoon side. The ocean-side beach is more shingly, with a higher proportion of pebbles and cobbles, whereas the lagoon-facing beach is generally sandier. There may be swampy depressions in between.

Both cays and motu may experience some degree of stabilisation by vegetation growth or by chemical lithification of their sediments. At the shoreline, mangroves trap sediment and inhibit longshore drift. On the island, vegetation colonisation helps to bind together the sediments with roots and provides a

---

[4] The word *motu* is used in both singular and plural forms.

layer of leaf litter that eventually forms a thin soil. Lithification of deposits takes place either within the beach to form a cemented layer of underlying beachrock, sometimes seen cropping out at the foot of the beaches, or 'inland' (if this concept has any meaning on a narrow motu or tiny cay) as aeolianite, sometimes called dune rock, or within the main body of the cay materials as cay sandstone (Stoddart 1971). Lithification may be aided on remote islands, where there are no human inhabitants, by the guano from colonies of roosting seabirds. The phosphate-rich guano reacts with the carbonate sediments, cementing them into a hard phosphatic rock.

7.4.2 Cay Migration and Redistribution

Overwash processes during tropical cyclones effectively strip small islets of their soil and fine sediments and may produce a combination of major changes. Geomorphic changes include the complete destruction of cays, reduction (or more rarely expansion) in cay size, changes in plan shape, cay migration across a reef or even relocation to entirely new positions on the reef platform. The depth to which an island is submerged is a governing factor. Whether or not islands have their natural vegetation intact with good undergrowth, or have been cleared and established with coconut plantations, also has a special influence on how much sediment stripping and adjustment in cay morphology takes place.

Unfortunately, little reliable historical data on cay size, shape or their exact position on reefs are available, which is necessary for studying their rates of episodic migration during storms. Yet there are reports of single powerful cyclones radically altering cay distribution. For instance, several new cays appeared on Ifaluk atoll in the Caroline islands after a typhoon in 1909, and Bayliss-Smith (1988) mentions several cays on Ontong Java atoll in Solomon Islands that either migrated, were reconstituted, or disappeared entirely as result of Tropical Cyclone Annie in 1967. Sometimes remnants of beachrock persist to mark the former outlines of lost or redistributed cays (Guilcher 1988).

7.4.3 Case Study – Geomorphic Change on Tatafa Cay in Tonga During Tropical Cyclone Isaac in March 1982

Tropical Cyclone Isaac in 1982 was one of the worst storms to affect the Kingdom of Tonga in the last century. The storm was relatively small, but traversed directly over the Ha'apai group of islands on 3 March with sustained winds of 148 km h^{-1} (80 knots). It then moved south west towards the main island of Tongatapu (Fig. 7.18), where pressure fell to 976.4 mb and 120 mm of rain was recorded in the capital Nuku'alofa (Woodroffe 1983). The arrival of the storm coincided with a high spring tide of 1.2 m and Nuku'alofa experienced extensive flooding to depths of several metres due to the storm surge.

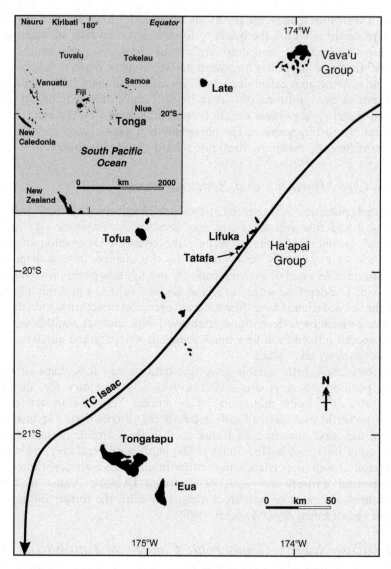

Fig. 7.18. Track of Tropical Cyclone Isaac across the Kingdom of Tonga on 3 March 1982.

Several small cays on the barrier reef north of Tongatapu were affected. Shorelines were washed out and reshaped. New exposures of beachrock occurred widely. The greatest changes in geomorphology were felt on the sand cays and beaches of the Ha'apai islands that lay in the path of the cyclone. The sand cay of Tatafa was particularly degraded. At its southern end, 200 m of the islet was completely removed by wave erosion. The sea swamped the island owing to the high storm surge, and powerful surge currents cut a new 16 m wide scour channel through the middle of the cay, which neatly bisected Tatafa into two smaller islets.

7.4.4 Atoll Hoa

Hoa are one of the most conspicuous geomorphic features of atolls in the (central) South Pacific. The term *hoa* is a Polynesian word, now commonly adopted in the literature on atolls, and is used to describe the shallow channels separating individual motu. Hoa[5] are considered to start from the lagoon side of an atoll and extend across the reef towards the oceanside. They are not to be confused with reef passes (gaps in the reef) known as *ava* in Polynesia. The characteristics of hoa vary considerably, even on the same atoll, and in consequence several attempts have been made at their classification into groups. The early work of Chevalier (1972) on the French Polynesian atolls is a benchmark in which the following six types of hoa were identified:

- Open on the lagoon side only – the most common type,
- Functional hoa that are open to the ocean and allow the exchange of seawater between the lagoon and ocean, either continually or perhaps only at higher stages of the tide or during storms,
- Open at the ocean side only, with the lagoon exit blocked with sediments or conglomerate,
- Open at the ocean side only, and closed by conglomerate at the lagoon exit,
- Blocked at both ends and enclosing a hypersaline channel,
- Dry hoa, either blocked and vegetated, or distinctly emerged above modern sea level. Emerged hoa are probably Holocene channels, abandoned by sea-level fall, and were termed palaeohoa by Stoddart and Fosberg (1994).

Over the last few decades, increasing attention has been directed towards the role of tropical cyclones as causative agents in hoa formation on Pacific atolls. Observations suggest that cyclones have several effects. Wave erosion may widen and deepen already active hoa. Unusual currents created by storm surge through the channels may enhance this process. New hoa may be eroded, either completely or perhaps only part way across a motu. Sometimes non-functional hoa (dry types or those open at one end only) may be cut through by large waves and thus reactivated into functional channels, for example at Rangiroa, Tikehau and Matavia atolls in the northwestern Tuamotu archipelago (Bourrouilh-Le Jan and Talandier 1985). Deltas or outwash fans may form at the mouths of hoa and extend into the atoll lagoon.

---

[5] The word *hoa* is used in both singular and plural forms.

Conversely, sediment deposition and the building of storm embankments may block ocean-side hoa exits. On Taiaro atoll (4.5°S, 172.2°W), Salvat *et al.* (1977) noted that non-functional hoa were associated with storm ridges 2 m high, and for Reao atoll (18.5°S, 13.4°W) in the Tuamotus, Pirazzoli *et al.* (1987) suggest that hoa on northern side of the atoll were blocked by a cyclone in 1903.

On Orona atoll (4.5°S, 172.2°W) in the southern Phoenix Islands of Kiribati, Stoddart and Fosberg (1994) became convinced that palaeohoa were cut by tropical cyclones during the Holocene at a time of high sea-level stand, by the sea overtopping the motu and eroding channels. The presence of large, basally eroded storm blocks sitting on the dry floor of the palaeo-hoa provided neat evidence for this idea on Orona. In addition, a wider view reveals that hoa are common on atolls within the cyclone belts, but rare on atolls beyond them. From this observation, Stoddart and Fosberg (1994) proposed that the existence of palaeohoa on the margins of modern cyclone-affected areas might be a useful clue for interpreting the spatial extension of Holocene palaeocyclone activity.

7.4.5 Motu Growth and Longevity

The supply of coral detritus by wave erosion of reef fronts during major tropical cyclones, and the associated depositional features of gravel sheets and rubble banks on reef surfaces, are arguably key processes necessary for the periodic growth and continued existence of low coral islands in the South Pacific. Almost a century ago, referring to atoll islands, Wood-Jones (1910, p. 261) proposed that:

Every detail of its [atoll island] structure points to the fact that increments have been added at quite irregular intervals: it seems as though the land had taken a rapid stride, and then has ceased from growing; the quiescent period again being followed by great additions to its substance. Sudden building increments are without doubt the evidence of storms of exceptional violence.

From observations in the atolls of the Marshall Islands, the Caroline Islands of the Federated States of Micronesia, and in the Tuamotu archipelago of French Polynesia, Newell and Bloom (1970) came to the conclusion that heights of atoll islands are more or less in equilibrium with present sea level, sediment supply and storm waves. Bayliss-Smith (1988) reviewed early studies both within and beyond the Pacific basin, and considered that catastrophic storms generating waves up to 10 m high are necessary for the production of large amounts of coarse coral debris. Such waves are the only mechanism capable of transporting this detritus from reef fronts onto reef surfaces.

After the construction of new rubble banks by powerful cyclones, the subsequent growth of motu then depends on longer-term coastal processes. These include the migration of the ramparts towards the shore, the reworking of this material by normal wave action, and the eventual nourishment of the coral islands by the bulk of these sediments. In the immediate post-cyclone period, the geomorphic disequilibrium is most pronounced and the heavy damage inflicted on reef-crest corals means that the energy of even medium-size waves cannot be dissipated. This gives a window of time for waves to reshape the rubble banks and effectively redistribute some of their loose deposits. As a result the banks may migrate landwards at rates between 1 and 10 m yr^{-1}, be spread out over the reef flat, or probably experience a combination of both. Over subsequent years, the heights of rampart crests decrease, the angles of the seaward-facing slopes reduce and the slope profiles change from convex to concave shapes (Baines and McLean 1976).

On Ontong Java atoll in Solomon Islands, Bayliss-Smith (1988) mapped the positions of large and extensive rubble banks constructed by TC Annie in November 1967 (Fig. 7.19). Field investigations showed that after 3 years the ramparts had been lowered from their original heights, and their constituent materials were largely reworked into more stable landforms by the action of lower-magnitude storms. On some islets the rampart formed a spit protecting the old shoreline, allowing the accretion of sand behind the spit. Elsewhere, the rubble banks were dispersed to form gravel sheets on the intertidal reef

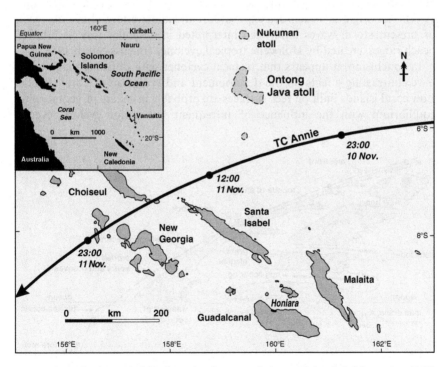

FIG. 7.19. Path of Tropical Cyclone Annie across Solomon Islands in November 1967, which caused geomorphic changes on many of the coral islands (motu) on Ontong Java atoll.

flat. TC Ida, which occurred in May 1972, was a less intense storm than TC Annie. It did not create any new rubble banks, but caused the 1967 rampart to join the original shores of islands in several places. Overall, the long-term geomorphic contribution of reworked cyclone-rampart debris included the:

- Production of low rubble sheets on reef flats in between individual islets,
- Addition of new increments to the seaward beaches of original shorelines,
- Development of spits extending some tens of metres beyond old shorelines.

Where multiple ridges are found stacked one against another (Fig. 7.20), this testifies to a past sequence of tropical cyclones. Stacked ridges may form a major component of the substance of windward motu, such as on Onotoa atoll (1.8°S, 175.6°E) in the Gilbert Islands of Kiribati (Cloud 1952).

Conglomerate platforms, which are benches of cemented coral rubble, are often observed between atolls islands and the outer edge of the reef flat on the ocean side. Newell and Bloom (1970) hypothesised that these benches represent the 'abraded roots' of old rubble banks created by ancient tropical cyclones (Fig. 7.21). Conglomerate platforms in Tuvalu and at Aitutaki in the Cook Islands are believed to have formed originally as storm ramparts, owing to their similar fabric composition, which became cemented and were eventually planed down by wave abrasion (Stoddart 1975, McLean and Hosking 1991). The origin of beachrock exposures found at elevations above the reach of present storm waves have been interpreted in a similar way as lithified beach ridges, created by Holocene tropical cyclones (Bayliss-Smith 1988).

In conclusion, it appears that tropical cyclones with differing magnitudes have contrasting effects on the development and evolution of coral islands. Low coral islands built on reef surfaces are probably in a state of geomorphic equilibrium with the influence of infrequent but major cyclone events.

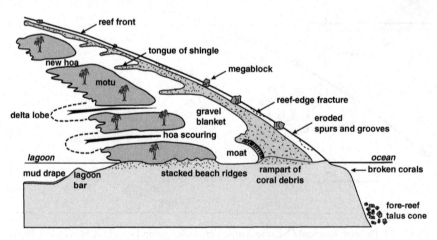

FIG. 7.20. Sketch of the main features of erosion and construction by a severe tropical cyclone on a South Pacific atoll. Adapted from Scoffin (1993).

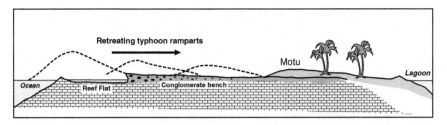

Fig. 7.21. An idealised profile across the rim of Raroia atoll in the Tuamotu archipelago of French Polynesia. The sketch illustrates the migration of rubble ramparts over the top of underlying conglomerate benches of cemented materials during quiescent periods between major cyclone events. Redrawn from Newell and Bloom (1970).

The role of episodic high-magnitude cyclones is to erode existing cays and motu, but at the same time provide the necessary coral debris for reconstruction, often built up into sizeable ramparts or laid out as thick gravel sheets. In between times, coral islands fluctuate in size, shape and position on the reef flat, and are in a state of geomorphic flux.

The task of lower-magnitude but more frequent storm events is to rework the material of existing rubble banks and use it to nourish the coast by rebuilding eroded shores and scoured beaches. This nourishment leads to the net growth of many coral islands. According to Bayliss-Smith (1988), on Ontong Java atoll this sequence appears to take about 20 years. Lithified Holocene-age cyclone ramparts may be planed down into conglomerate platforms overlying reef surfaces, and then form suitable basement foundations for the accretion of modern cays and motu.

7.4.6 Case Study – New Land Created on Funafuti Atoll by Tropical Cyclone Bebe in October 1972

Funafuti atoll (8.5°S, 179.2°E) in the atoll nation of Tuvalu was struck by Tropical Cyclone Bebe on 21 October 1972. TC Bebe approached initially from the east, but the track then followed an unexpected clockwise loop near the southeast coast of the atoll before curving away southwards. This unusual motion led to a high storm surge at Funafuti, and a monstrous storm wave consequently swept across the eastern half of the atoll. The impact on the coastal geomorphology was the construction of an enormous rampart of coral debris (Table 7.2).

The rampart was built on top of the ocean-side reef flat, and stretched almost continuously 18 km along the southeastern perimeter of the atoll (Fig. 7.22). The rampart was deposited well above the intertidal zone and was nearly as large as many of the pre-existing islands on the atoll (Maragos *et al.* 1973). Poorly sorted coral rubble and shingle made up the majority of the fragments, within which the skeletal remains of recently living *Acropora*,

TABLE 7.2. Dimensions of the rubble rampart built on Funafuti atoll in Tuvalu by Tropical Cyclone Bebe on 21 October 1972.

Rampart features	Dimension
Length	18 km
Height (mean)	3.5 m
Width (mean)	37 m
Total volume	$1.4 \times 10^6 \text{ m}^3$
Rubble mass	2.8×10^6 tons
Size of surface clasts (mean)	9–10 cm
Diameter of largest block observed	7 m
Width of back-rampart moat	2–50 m

From Maragos *et al.* (1973).

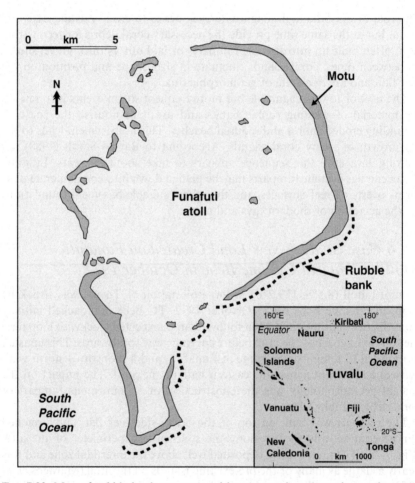

FIG. 7.22. Map of rubble banks constructed by wave action along the southeastern reefs of Funafuti atoll in Tuvalu by Tropical Cyclone Bebe on 21 October 1972. Redrawn from Maragos *et al.* (1973).

FIG. 7.23. Coarse beach gravels on Funafuti atoll in Tuvalu. Beaches have been nourished by the degraded remains of rubble ramparts created by Tropical Cyclone Bebe in October 1972, which subsequently migrated through wave action across the reef flat to join ocean-facing shores of the main islands. Photo by Randy Thaman.

Pocillopora and *Pavona* corals could be identified. Over subsequent years the ramparts migrated towards, and eventually merged with, the shores of the main motu (Fig. 7.23). The long-term effect of TC Bebe has therefore been to add a significant quantity of coarse sediment to the substance of Funafuti's coral islands, undoubtedly aiding their longevity.

Chapter 8
Slope Stability and Mass Movements

8.1 Slope Susceptibility to Failure

The rugged terrain of the high volcanic islands in the South Pacific is susceptible to landslides, debris flows and other types of mass movements. Many extensive slope failures are activated during severe tropical cyclones.

Slope susceptibility to failure is influenced by the predominance of clay-rich soils, normally humic latosols, overlying residual red/orange saprolite. Saprolite is completely-weathered rock and sediments *in situ*. The *regolith* (both soil[1] and saprolite material) is formed by chemical weathering of the common types of volcanic rocks and volcanic-derived sediments found in the South Pacific islands, especially andesites and basalts, and associated breccias and conglomerates. The regolith often has structural weaknesses inherited from the bedrock and overlying sediments from which it is derived, such as inclined bedding planes, faults and sediment/bedrock junctions. These zones of weakness may develop into shear planes along which the regolith mass fails in a landslide. If a particular area has a history of past landslides, it means that hillslopes are formed on weak residual materials and are prone to further failures.

Yet under normal weather conditions, including rain events, regolith materials in the tropical Pacific islands are capable of supporting steep slopes without failing. This is due to their beneficial hydrological properties, especially high soil porosity and permeability. This means that soil-moisture throughflow (and more rarely occurring natural soil pipeflow) is consequently an effective hillslope-hydrological process. The result is that volcanic soils and regolith can usually accept high rainfalls without saturating, owing to their rapid transmission of moisture down slope. Soil cohesion does not decrease quickly during average precipitation episodes, because critical pore-water pressure is not developed in the pore spaces and soil strength is maintained. Typical values of

---

[1] Many slope geomorphologists and engineers use the terms soil and regolith interchangeably.

TABLE 8.1. Shear strength properties in deeply-weathered volcanic regoliths at 11 sites along the main highway in the wet zone of southern Viti Levu island, Fiji. Parent rock types are mostly andesites and related breccias, conglomerates and sandstones.

	Consolidated drained tests		Consolidated undrained tests				
	Cohesion (kPa)[a]	Internal angle of friction (°)	Cohesion (kPa)	Internal angle of friction (°)	Peak shear strength (kPa)	Residual shear strength (kPa)	Sensitivity[b]
Average	45.2	25.8	49.8	24.5	104.0	29.8	3.5
Range	7–83	14–32	4–110	11–37	39–169	16–47	2.5–4.3

Summarised from Lawson (1993).
[a] $1 \text{ kPa} = 10.197 \times 10^{-3} \text{ kg cm}^2$.
[b] Sensitivity is the ratio between undisturbed and disturbed shear strength.

soil cohesion and strength in highly weathered regolith derived from volcanic rocks in southern Fiji is given in Table 8.1.

When landslides do occur, *shear failure* within the regolith is normally responsible, or in the case of debris flows at least for their initiation. Shear failure takes place when the available shear strength (s) of the slope materials is exceeded by the *in situ* shear stress (τ). Shear stress is the stress that acts along (parallel to) a plane on which a force has been applied. Shear strength is the internal resistance of a material against shear stress.

Shear strength is defined by the modified Coulomb–Terzaghi model (Rouse and Reading 1985), which can be expressed mathematically as:

$$s = c + (\sigma - \mu)\tan \varphi,$$

where s is the shear strength, c the cohesion, σ the normal stress (on the failure surface), μ the pore-water pressure and φ is the internal angle of friction (angle of shearing resistance).

Shear failure is brought about either by a reduction in shear strength, or by an increase in shear stress due to static and dynamic loads (Greenbaum *et al.* 1995). From the equation above is seen that shear strength will decrease when there are adjustments in the parameters on the right-hand side of the equation, namely if either:

- Soil cohesion weakens (c decreases), and
- Pore-water pressures are enhanced (μ increases, so the term $(\sigma - \mu)$ decreases).

8.2 Failure Trigger Mechanisms

A *trigger mechanism* is a discrete, identifiable event or influence that initiates slope failure. Tropical cyclones and earthquakes are both very effective trigger mechanisms. Tropical cyclones are able to provide the necessary trigger for slope failure in two different ways.

First, dynamic loads are increased through the violent swaying of tall trees during cyclone-force winds, because stresses are transferred to the slope materials by agitation of the root networks. A good example of this occurred in Samoa during the passage of Tropical Cyclone Val between 6 and 10 December 1991. TC Val was a hurricane-intensity system that did considerable damage to rainforests across the country. On the island of Upolu in the basin of the main Vaisigano river, a post-cyclone damage assessment showed that 40% of trees were blown down (Baisyet 1993). Uprooting was significantly more frequent among tall, large diameter trees, and species lacking buttresses or stilt roots were worst affected (Elmqvist et al. 1994). Numerous landslides were initiated by the uprooting of trees, concentrated in particular in the central and southern parts of the Vaisigano catchment (Fig. 8.1).

Second, exceedingly high rainfalls are required before pore-water pressures in the regolith are sufficiently elevated to induce failure, but tropical cyclones are able to deliver the necessary sustained rainfalls at exceptional intensities (Fig. 8.2). Precipitation with these characteristics exceeds the maximum capacities for infiltration and subsurface evacuation of soil moisture by throughflow, and the soil matrix eventually becomes saturated. Critical pore-water pressures are then reached, whereupon tropical clays lose their structure and cohesion, with a corresponding reduction in soil shear strength and the activation of mass movements.

This type of process triggered numerous slope failures on Vanua Levu island in Fiji on 7 March 1997 during Tropical Cyclone Gavin. In one example that nearly resulted in tragedy, a prolonged rain deluge over the north coast of the island as the cyclone passed nearby caused a large landslide that destroyed a residential dormitory for 65 children at a rural school in the village of Nabala. Fortunately the children were absent as the day was a national public holiday. According to investigations by the Fiji Mineral Resources Department (MRD pers. comm.), poor drainage in the upper portion of the slope led to soil saturation and surface ponding of water during TC Gavin. This set off a series of rotational slips in deep soil layers along an antecedent fault zone of structural weakness. After initial failure, the central rotational slip subsequently underwent liquefaction, which turned it into a rapidly moving mudflow that travelled over 80 m down slope and engulfed the school buildings (Terry and Raj 1999).

Mass movements induced by torrential rainfall may occur on slopes of widely varying slope angle, in both forested and non-forested areas. In early unpublished reports of landslides in Fiji, reviewed by Greenbaum et al. (1995), the majority of failures were recorded as starting from upper sections of hillsides, at or near slope crests. This indicates that sliding in many cases was caused by downward percolation of rainwater rather than by rising

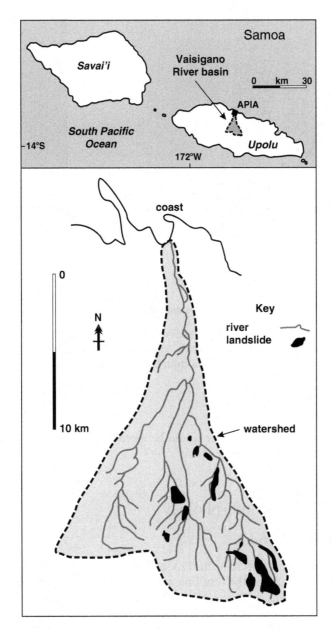

FIG. 8.1. Landslides initiated mainly by tree uprooting in the Vaisigano river basin on Upolu island in Samoa. Tree-throw was caused by hurricane-force winds during Tropical Cyclone Val in early December 1991. Data from Baisyet (1993).

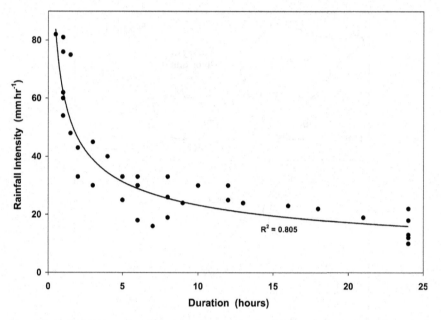

F𝘪𝘨. 8.2. Rainfall intensity–duration relationship for 35 storms including 16 tropical cyclones between 1979 and 1990, which caused slope failures in the highlands of southern Viti Levu island in Fiji. Data from Lawson (1993).

groundwater tables, as the latter would be more likely to influence the lower slopes first and to the greatest extent.

8.3 Landslide Geomorphology

8.3.1 Major Features

Whatever the trigger mechanism responsible, once slope failure is initiated, the shape and nature of the resulting mass movement is largely governed by the composition and thickness of the slope-forming materials and the presence of pre-existing weaknesses or discontinuities in the slope. *Debris flows* and *debris slides* (Fig. 8.3) tend to be very common types of failure in the mountains and hilly terrain of volcanic islands in the South Pacific. Debris flows are composed of large clasts carried by a mud and water mixture. Flows move as a ruptured mass, and the resulting deposits are poorly sorted and often internally structureless. A debris slide is a shallow type of mass wasting, moving by lateral displacement along a relatively narrow failure zone. Debris slides are characterised by shearing and plastic deformation along single or

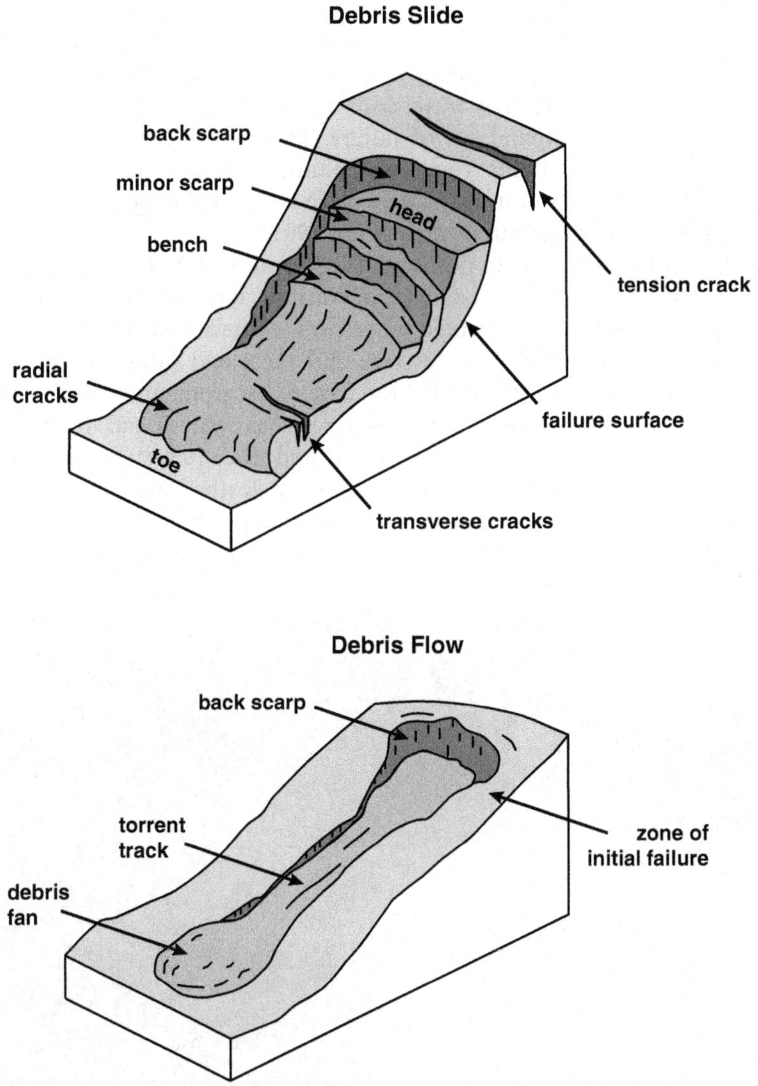

Debris Slide

Debris Flow

FIG. 8.3. Debris slides and debris flows, the most common types of mass movement on steep slopes of volcanic islands in the tropical South Pacific.

composite failure planes, such as joints and bedding within a weakened soil mass. The location of individual landslides is influenced by spatial variations in the nature of the surface regolith (Howorth *et al.* 1981), such as thickness, degree of weathering, and content of lithic fragments.

Debris flows and debris slides are often confined to the top 2–5 m of residual soils and weathered rock (Lawson 1993), although more rarely they can involve the complete regolith cover down to unweathered bedrock. They may erode areas exceeding a hectare in size and are sometimes capable of transporting enormous boulders up to 25 m in diameter into rivers (Trustrum *et al.* 1989). Debris flows may travel considerable distances and be deposited well away from the original site of failure. This is because although debris flows are initiated by sliding, they quickly transform into viscous yet highly mobile slurries. Deposition only takes place when the slope gradient is no longer sufficient to maintain flow (Lawson 1991).

TC Gavin in early March 1997 and TC June in mid-May 1997 generated some notable mass wasting activity of various types in the Fiji Islands. Overflight viewing of the volcanic ranges in the interior of Viti Levu island revealed that many rockfalls from steep cliff faces had subsequently set off debris slides in talus accumulated at the base of escarpments. Many of these shallow translational slides with near-surface shear planes had driven paths through the rainforest vegetation (Terry and Raj 1999). Media reports described other large failures that caused several deaths, damaged infrastructure or destroyed human habitations in and around urban areas (Fig. 8.4).

FIG. 8.4. A shallow translational slide, approximately 40 m wide, below Edinburgh Drive in Suva City, the capital of Fiji. Following several days of heavy rainfall in advance of Tropical Cyclone June in early May 1997, this failure in very friable and highly weathered soapstone (Suva Marl) removed one lane of the main road. Photo courtesy of the Fiji Times.

Deep-seated landslides are mass movements that fail along shear planes that are deeper beneath the surface. They normally occur on regolith that is thicker than average, for instance in the wettest locations on the windward sides of tropical volcanic islands, or at high elevations in the mountains, often under tropical rainforest. In such locations pronounced chemical weathering of the bedrock produces very thick layers of clay-rich saprolite beneath the soil.

Deep-seated failures are not as common as shallow failures, but their occurrence is not unknown during intense tropical cyclones. From 18 to 19 May 1986, the track of Tropical Cyclone Namu passed through Solomon Islands (Fig. 8.5), traversing the islands of Malaita and Guadalcanal. TC Namu inflicted terrible destruction and shaped considerable landscape change, including numerous debris flows and deep-seated landslides. The enormous amounts of debris choked valley bottoms, leading to unprecedented river channel aggradation on both Malaita and Guadalcanal (described in Chapter 10). At least one hundred lives were lost, primarily as a result of landslides in the hills and highlands (Trustrum *et al.* 1990). Table 8.2 shows the number of deep-seated landslides in affected river basins in Solomon Islands. The data was observed from 119 aerial photographs

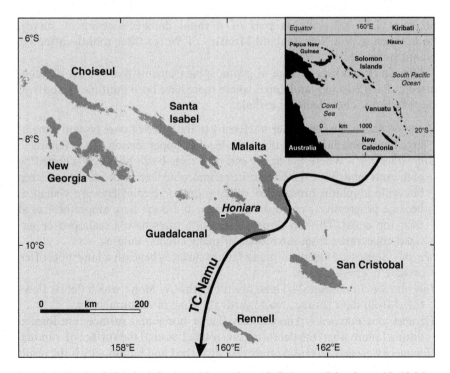

Fig. 8.5. Track of Tropical Cyclone Namu through Solomon Islands on 18–19 May 1986, which triggered numerous deep-seated landslides (see Table 8.2).

TABLE 8.2. Deep-seated landslides on Guadalcanal and Malaita islands in Solomon Islands, which were triggered by Tropical Cyclone Namu on 18–19 May 1986.

River basin	Basin area (km^2)	Number of deep-seated landslides
Guadalcanal Island		
Mbalisuna	235	42
Ngalimbiu	262	37
Mberande	234	24
Mbokokimbo	385	15
Nggurambusa	205	9
Tavangaoa	173	7
Lughumboko	216	5
Matepono	202	4
Rere	109	3
Kolovaghamela	88	2
Malaita island		
Kwariakawa	78	15
Wairaha	486	14
Kwaleungga	364	10
Kwaimbaita	195	3
Waitahu	1	1

Data from Stephens *et al.* (1986).

taken at 1:25,000 scale as part of a rapid damage-assessment survey, undertaken by the New Zealand Ministry of Works three months after TC Namu struck.

Lawson (1993) lists a range of geomorphic elements that are easily recognised in Pacific island landscapes where there have been multiple episodes of slope failures. These features include:

Scars – A scar is the general term referring to the patch of bare ground remaining after a landslide, normally at the head or upper section of the failure.

Amphitheatres – Where the upper end of a river basin has been sculpted by numerous landslides, the composite geomorphic feature produced is a steep but wide amphitheatre-headed valley (normal river valleys are v-shaped, become progressively narrower upstream, and don't have amphitheatres at their top ends). The top slopes of amphitheatres have a scalloped or serrated appearance from the mosaic of many circular failures.

Scarps – Scarps are steeply dipping failure surfaces beneath a landslide, often arcuate in cross-section.

Torrent tracks or *Chutes* – These are the pathways along which debris flows travel from their source area towards the zone of accumulation.

Terraces and *Benches* – Anomalously flat or horizontal surfaces are formed during failure when the debris mass is rotated around the surface of rupture.

Hummocky ground – Terrain appearing disturbed and hummocky is the result of disruption within the regolith mass during failure.

Fans – At the distal end of debris flows, the accumulation of deposits is often fan-shaped in plan.

8.3.2 Case Study – the 'Good Friday Landslides' Triggered on Viti Levu Island by Tropical Cyclone Wally in April 1980

The effects of Tropical Cyclone Wally on landslide activity in the southern part of Fiji were investigated shortly after the event by Crozier *et al.* (1981) and Howorth and Prasad (1981). TC Wally bears the distinction of being one of very few cyclones in recorded history to make a complete traverse of Viti Levu island. The track crossed the island directly from north to south over the Easter period of 1980 (Fig. 8.6). A spectacular amount of mass wasting

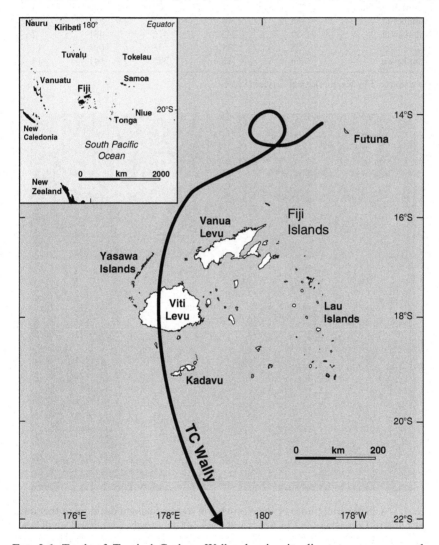

FIG. 8.6. Track of Tropical Cyclone Wally, showing its direct traverse across the mountainous volcanic island of Viti Levu in Fiji on 4 April (Good Friday) 1980. Numerous hillsides failed in response.

was triggered, owing to the immense downpour of rain over the southeast quadrant of Viti Levu (Table 8.3). Innumerable hillslopes failed throughout the Serua and Navua areas (Fig. 8.7). Many of these were massive failures. The enormous quantities of transported debris compounded the already

TABLE 8.3. Daily precipitation measurements at selected climate stations in southern Viti Levu, Fiji, during Tropical Cyclone Wally at the start of April 1980.

Climate station	Rainfall (mm)				
	1 April	2 April	3 April	4 April	5 April
Laucala Bay (Suva)	32	111	308	262	14
Wainibora	45	224	505	552	9
Nabukavesi	50	243	416	481	7
Sakisa	55	271	538	601	17
Wainikavou	66	158	320	641	14

Data source: Fiji Meteorological Service (1980).

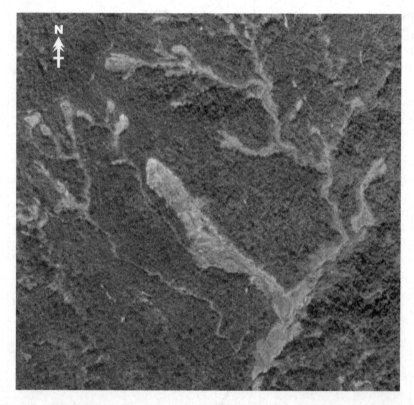

FIG. 8.7. A 1990 aerial photo of the Korovisilou river catchment draining to the south coast of Viti Levu island in Fiji. Scars and tracks of a number of large and small landslides are seen. At the time of the photo most landslides remained poorly vegetated, even though 10 years had already elapsed after their original failure caused by Tropical Cyclone Wally on 4 April 1980. A recent photograph of the large and elongate debris flow in the centre of the image is shown in Fig. 8.8.

FIG. 8.8. Path of an old debris flow that continues to be unstable, on the slopes of the Korovisilou river catchment in southern Viti Levu, Fiji. The original failure was initiated by intense rain during Tropical Cyclone Wally on 4 April 1980. Photo by Michael Bonte in 2005.

extensive devastation along Viti Levu's south coast, which was caused by most coastal rivers and streams bursting their banks.

One of the most serious effects of landslides in the Serua Hills area to the west of Navua town was disruption to communications along the Queen's Road highway. This is the arterial route through the southern part of the island. In many places the road was either blocked by sediments, or undermined and eroded. Rural communities faced isolation for up to four weeks (Howorth *et al.* 1981). When anecdotal accounts provided by local Fijian villagers was compared, the consistent story emerged that most landslips occurred on the evening of 4 April 1980, because the ground became saturated after heavy and continuous rain over the previous two days. This day was Good Friday at the start of the Easter weekend, so the episode is remembered as the 'Good Friday landslides'.

Some of the larger slope failures set off by TC Wally are still active today, 26 years later. This means that the original debris flow deposits have not yet stabilised, but continue to move slowly down slope by soil creep processes. This prevents rainforest from being re-established, with only low-level scrubby vegetation managing to grow over the surface of the deposits.

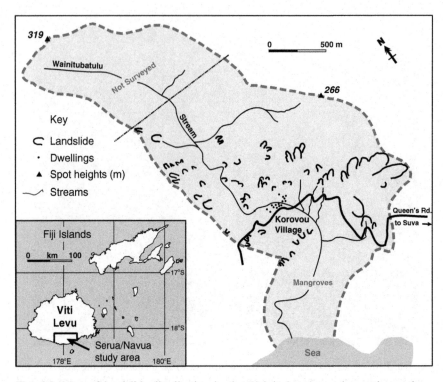

FIG. 8.9. Map of landslide distribution in the Wainitubatolu catchment in southern Viti Levu, Fiji, caused by Tropical Cyclone Wally on 4 April 1980. Simplified from original in Crozier *et al.* (1981).

Consequently the scars and torrent tracks of the 1980 landslides are still clearly visible, as seen in Fig. 8.8.

Shortly after the TC Wally event, Crozier *et al.* (1981) investigated the steep and dissected 313 ha coastal catchment of the Wainitubatolu stream that drains into Korovou Bay on Viti Levu's southern coast (Fig. 8.9). This catchment was chosen primarily because it was one of the areas most severely affected by landslide activity. Observations in the field identified 74 separate mass movements. Most were described as debris flows. However, on especially steep or gentle slopes, respectively, some failures were noted instead to show characteristics suggesting processes more akin to avalanching or sliding during failure.

The majority of debris flows formed in surfical regolith 0.3–10 m thick, with failure planes invariably located along the regolith/bedrock contact. Debris flows originated in upper slope locations, or in ridge and summit concavities, but the fluid mass travelled down slope as far as the valley floor. Many flows entered drainage networks, choking stream channels with a mixture of mud and coarse sediments. At the head of some larger landslides, well-defined amphitheatre-shaped scars exposed the underlying volcanic

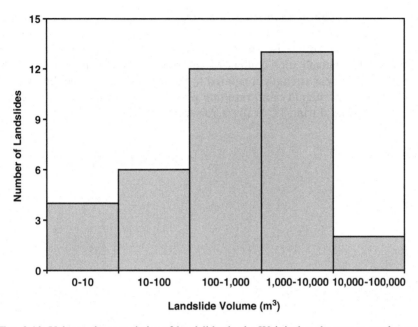

FIG. 8.10. Volume characteristics of landslides in the Wainitubatolu stream catchment in southern Viti Levu, Fiji, caused by Tropical Cyclone Wally on 4 April 1980. Replotted from data in Crozier *et al.* (1981).

TABLE 8.4. Landslide activity triggered by Tropical Cyclone Wally in April 1980, on two contrasting types of regolith in the Wainitubatolu catchment in southern Viti Levu island, Fiji.

Measured variable	Red/orange clay regolith	Brown granular regolith
Hillslope area (ha)	135	79
Number of slips	62	12
Percentage of all slips (%)	84	16
Slip density (slips km^{-2})	46	15
Volume of material displaced (m^3)	103,424	5,858
Displacement ratio (m^3 ha^{-1})	766	74
Percentage of hillslope area slipped (%)	1.9	0.8

Recalculated and simplified from Crozier *et al.* (1981).

bedrock. Overall it was apparent that localised perching of the water table had caused widespread rupture of the regolith. The intense and prolonged rainfall provided by TC Wally was responsible for this, combined with the concentration of water on slopes by the geomorphic configuration of slope concavities. Size characteristics of the debris flows are shown in Fig. 8.10.

One of the main findings was that the spatial pattern of slope stability and instability had been influenced by the distribution of two distinctive types of regolith (Table 8.4). On deeply weathered red regolith dominated by red/orange clay, landslides were observed to be significantly deeper and three times more common than on slope areas underlain by brown granular regolith. The brown granular regolith is less weathered and therefore possibly younger, and is characterised by enhanced internal friction and permeability owing to its relatively high content of volcanic rock fragments. This affords it with higher shear strength compared to the weaker red clay regolith, which influenced slope stability and resulting patterns of landslide activity across the steep terrain during TC Wally (Crozier *et al.* 1981).

Chapter 9
River Hydrology and Floods

The rivers overflow their banks, and water covers the plains; uprooted trees, wrecked buildings, and drowned cattle sweep by on the tawny flood. So the storm rages for hours.
(Derrick 1951, p. 115)

9.1 Introduction to Island Rivers

The geology of the volcanic islands in the South Pacific generally consists of lava flows, pyroclastics, breccias and conglomerates, all weathered to varying degrees according to the age of the individual island concerned. Geomorphology is often dominated by volcanic mountains forming a central highland area, such as on Rarotonga in the Cook Islands, Viti Levu in Fiji, Ambrym in Vanuatu and Tahiti in French Polynesia. Orientation of the major drainage networks tends to be in a radial fashion outward from the central highlands. If the volcanoes are aligned in a chain, then the volcanic peaks form an elongated mountainous spine to the island. Examples include Savai'i in Samoa, Kadavu in Fiji, Santa Isabel in Solomon Islands and Pentecost in Vanuatu. On such islands, the orientation of the major river networks is controlled by the linear arrangement of the volcanic mountains. Within individual catchments, drainage patterns are typically dendritic because of the lack of geological or structural controls other than the volcanoes. Drainage densities are high. Centripetal drainage patterns are seen where a river drains a breached volcanic caldera, such as the Tavua and Lovoni rivers on the islands of Viti Levu and Ovalau in Fiji, respectively.

On smaller South Pacific islands, the main watercourses are streams with catchment sizes of perhaps several tens of square kilometres in area. Only the larger islands have river basins extending across several hundred square kilometres (see Table 9.1). At their headwaters, river watersheds tend to be separated by narrow, serrated interfluves. Slope angles are steep, often approaching 30° or more. The upper reaches of river courses tend to be steep and bouldery (Fig. 9.1), sinuous and with long-profiles dominated by step-pool architecture. Coarse bedload is normally abundant. Short sections of bedrock channel and waterfalls occur frequently where resistant lava flows have been

TABLE 9.1. Largest rivers on selected islands in the South Pacific.

Island	Country	Largest river	Approximate basin area (km^2)
Viti Levu	Fiji	Rewa	2,918
Malaita	Solomon Islands	Wairaha	486
Grande Terre	New Caledonia	Yate	437
Guadalcanal	Solomon Islands	Lungga	394
Espiritu Santo	Vanuatu	Jourdain	369
Vanua Levu	Fiji	Dreketi	317
Savai'i	Samoa	Sili	51
Upolu	Samoa	Vaisigano	33

FIG. 9.1. Bouldery channel of an upland tributary of the Navua River, southern Viti Levu island in Fiji.

exposed, especially on islands experiencing tectonic uplift. Stream banks of upland tributaries are steep, composed of bouldery soils and regolith, obscured by dense grassland or rainforest. In first- and second-order streams, no obvious boundary is visible between stream banks and the steep valley-side slopes. Rock outcrops near slope summits create near-vertical sections to many valley cross-profiles, especially where valleys are guided by faults.

Deep river downcutting resulting from high annual precipitation, often accelerated by uplift, has produced a highly dissected fluvial landscape on many islands, and some impressive gorges are seen along some of the larger rivers (Terry *et al.* 2002a). The Navua, Namosi and Waiqa gorges on Viti Levu island in Fiji are fine examples. Gorges are sometimes associated with amphitheatre-headed valleys, for instance the Wabu valley on Viti Levu and the Ngatoe valley in southern Rarotonga in the Cook Islands (Nunn 1994). Physical geographic factors necessary for amphitheatre development are volcanic relief rising above 500 m in elevation and annual rainfall in excess of 2,000 mm. This allows rapid fluvial incision in the highlands (compared to lower slopes where orographic rainfall is less) and simultaneous deep chemical weathering on the sheer valley sides. Thick saprolite is produced that is prone to mass wasting during prolonged heavy rain.

Lower sections of river basins generally have more subdued (hilly) terrain, with flat alluvial terraces and floodplains, or more rarely braidplains, in valley bottoms (Fig. 9.2). Rivers debouch onto narrow coastal plains of recent colluvium and alluvium, although examples of watercourses emptying over waterfalls directly into the sea are not so unusual, for example the Falefa and

FIG. 9.2. Meandering reach of the lower Waidina River, eastern Viti Levu island in Fiji.

Savulevu rivers on Upolu island in Samoa and Taveuni island in Fiji, respectively. The bigger rivers in Solomon Islands, Vanuatu, New Caledonia and Fiji have built expansive deltas. Floodplain and deltaic areas are normally farmed or prized for cattle grazing, and have relatively high population densities compared to upland interiors of volcanic islands. This makes modern island populations vulnerable to river floods.

9.2 Tropical Cyclone Floods

The huge and intense rainfalls brought by tropical cyclones generate very quick responses in Pacific island streams and rivers. Peakflows are often far in excess of maximum channel capacities, leading to severe overbank inundation (Fig. 9.3). River floods are a serious hydrological hazard because of their impacts across both physical and human landscapes. Flood problems include channel erosion and siltation, destruction of homes and infrastructure, contamination of water resources, damage to subsistence agriculture (threatening food security) and obvious risks to human life and health. Tropical cyclone-induced river flooding normally causes more fatalities and general devastation than the combined effects of violent winds, large waves and powerful storm surges described in earlier chapters.

The short- and long-term consequences of river floods can place heavy socio-economic burdens on developing island nations (Table 9.2, Terry 2005).

FIG. 9.3. Flooding of the lower Lungga River near Honiara on Guadalcanal island in Solomon Islands, on 18 May 1986 during Tropical Cyclone Namu. Courtesy of the Solomon Islands Department of Mines and Energy.

TABLE 9.2. Costs of flood damages in Fiji caused by selected tropical cyclones in recent decades.

Tropical Cyclone	Year	Damages (Fiji$ million[a])
Oscar	1983	148
Eric	1985	64
Gavin	1985	2
Sina	1990	33
Joni	1992	2
Kina	1993	188
Gavin	1997	35
June	1997	1

From Feresi *et al.* (1999).
[a]Converted to 1998 Fiji$ values.

Yet in spite of the problems they cause, the characteristics of cyclone-related floods are generally poorly understood in the South Pacific (Kostaschuk *et al.* 2001). Improved mitigation of flood disasters has been identified as a priority by many island nations in recent decades. However, this requires a better understanding of the hydrological behaviour and response of island fluvial systems to tropical cyclones than currently exists. For the purpose of this chapter, the term *flood* means the peak river flow (or *peakflow*) during a tropical cyclone, which may or may not cause inundation, whereas *overbank flood* refers only to flows that exceed the capacity of the channel banks (known as the *bankfull discharge*) and actually cause floodwaters to spill out over the land.

9.2.1 Influences on River Responses

High-magnitude rainfalls delivered by tropical cyclones normally produce extraordinarily big discharges in Pacific island rivers (Figs. 9.4 and 9.5). The nature of a river's response depends primarily on three groups of influences, which are discussed below. The first group are the meteorological characteristics of an individual tropical cyclone, such as the organisation of its cloud bands, speed of movement and the corresponding patterns in precipitation. The second set of factors are those concerning the physical geography of the landscape, particularly the types of geology, soils, topography and vegetation. The third group are parameters related to the geometric configuration of the river drainage basin, especially its size, shape and orientation.

During storm events, it is the combination of exceptional rainfall amounts and intensities that promotes swift transmission of moisture into stream channels, by saturating the soil and generating excess runoff (Bonell and Gilmour 1978, Walsh 1980). Contrary to popular belief, this can occur under a mature cover of tropical rainforest (Herwitz 1986). Storm duration is also important because rainfall interception by the vegetation canopy decreases in proportion to storm size (Jackson 1971). This means that the longer a storm lasts, the more the canopy interception decreases and throughfall increases as a percentage of total precipitation input. Windy conditions in tropical

FIG. 9.4. View of the old main bridge over the Sigatoka River at Sigatoka town, southwest Viti Levu island in Fiji, during the flood produced by Tropical Cyclone Kina in December 1992. The water level is 10 m or more below the bridge at normal flow. Courtesy of the Fiji Times.

cyclones also aid moisture transfer to soils by further reducing the possibilities for vegetation interception (Herwitz 1985). If two or more cyclones strike an island during the same wet season, then forest damage such as foliage stripping, crown breakage and tree uprooting caused by the initial storm reduces the interception-storage capacities of the canopy during any later cyclone events (Walsh 1982), encouraging very rapid water loss from affected catchments.

The windward side of an island in relation to the direction of cyclone approach, which we might call the 'cyclone side', will benefit from orographic effects and generally receive much more rainfall than the sheltered side behind volcanic highlands. Consequently, rivers on the cyclone side of islands are more likely to experience severe floods. For Vanuatu, Samoa and Fiji, rivers on the north and west sides of the big islands are therefore more prone to cyclone-induced floods. This is because in addition to draining tall mountains that promote strong orographic effects, they also face the direction from which tropical cyclones normally arrive.

The speed and the proximity of a cyclone track in relation to the position and orientation of a particular river basin are also important influences on the flood magnitude. This is because a storm which travels slowly, close and roughly parallel to the long axis of a river basin, is likely to deliver big rainfalls to most of the river's main tributaries and thus produces a large flood. Flooding is less likely where a cyclone moves relatively quickly along a track that runs perpendicular to the orientation of the river system (Kostaschuk

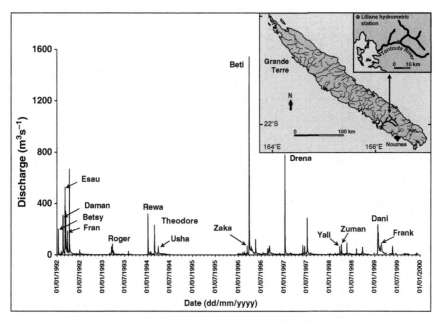

FIG. 9.5. Peakflows produced by tropical cyclones from 1992 to 1999 in the Tontouta River on Grande Terre island, New Caledonia.[1] Bankfull discharge at the hydrometric station is approximately 600 m^3 s^{-1}. Source: Observatoire de la Ressource en Eau, New Caledonia.

et al. 2001). This is illustrated by Table 9.3, which lists the chronology of tropical cyclones that have caused overbank floods at three hydrometric stations in the major Rewa River system in Fiji (Fig. 9.6) between 1970 and 1997. Tropical cyclones that had the greatest flood impacts were those that passed near the axis of the watershed. The centres of TCs Wally, Bebe and Kina, for example, all tracked within 50 km of the centre of the Rewa basin, and more importantly followed paths that ran more or less in the same direction as the basin orientation.

Since many volcanic island rivers have upper catchments dominated by rugged topography, this promotes a high degree of *hydrological short-circuiting*. This means that during a torrential cyclonic downpour, the precipitation is transferred quickly into the river channels, leading to a fast hydrological response. Rivers therefore display *flashy* behaviour, giving little time lag

---

[1] Grande Terre island in New Caledonia is not a volcanic island like the large islands of neighbouring Solomon Islands, Vanuatu and Fiji. It is a piece of the ancient continent of Gondwanaland and comprises ultramafic rock types. Grande Terre has steep terrain and good flow records exist for the Tontouta River. So the Tontouta is a useful river for illustrating the effects of tropical cyclones on river hydrology in the South Pacific, in spite of the unusual (non-volcanic) basement geology.

TABLE 9.3. Chronology of tropical cyclones from 1970 to 1997 that caused overbank floods at three hydrometric stations in the Rewa River system of eastern Viti Levu island in Fiji.

Tropical cyclone	Storm duration (dd/mm/yy)	Nabukaluka		Nairukuruku		Navolau	
		Flood duration (days)	Q_p (m³ s⁻¹)	Flood duration (days)	Q_p (m³ s⁻¹)	Flood duration (days)	Q_p (m³ s⁻¹)
Priscilla	14/12/70–18/12/70	<	<	–	–	1	2,003
Bebe	19/10/72–06/11/72	1	–	–	–	3	6,711
Lottie	05/12/73–12/12/73	<	<	–	–	2	2,304
Tina	24/04/74–28/04/74	<	<	–	–	1	3,100
Val	29/01/75–05/02/75	<	<	–	–	2	2,806
Wally	01/04/80–06/04/80	2	798	2	1,853	2	4,218
Hettie	27/01/82–30/01/82	<	<	2	1,609	2	3,311
Oscar	28/02/83–02/03/83	<	<	2	2,759	3	4,533
Nigel	19/01/85–20/01/85	<	<	2	1,368	1	2,734
Gavin	04/03/85–07/03/85	<	<	2	2,140	2	3,960
Sina	24/11/90–30/11/90	<	<	<	<	1	2,117
Joni	06/12/92–13/12/92	<	<	1	2,089	2	3,581
Kina	26/12/92–05/01/93	2	602	4	7,334	2	6,923
Gavin	04/03/97–11/03/97	<	<	3	6,384	1	3,927
June	03/05/97–05/05/97	<	<	–	–	3	4,015
Flood characteristics							
Number of overbank floods			2		7		15
Q_p sample mean (m³ s⁻¹)			700		3,192		3,750
Q_p standard deviation (m³ s⁻¹)			–		2,314		1,469
Decadal frequency of overbank floods			1.1		2.9		5.4

Storm duration is the time that the tropical cyclone occupied Fijian waters. Q_p is the peak daily discharge caused by the cyclone, and flood duration is the period that the flow exceeded bankfull stage; < indicates the peakflow remained below bankfull discharge (i.e. no overbank flood), – indicates no data. Source: Hydrology Division of the Fiji Public Works Department.

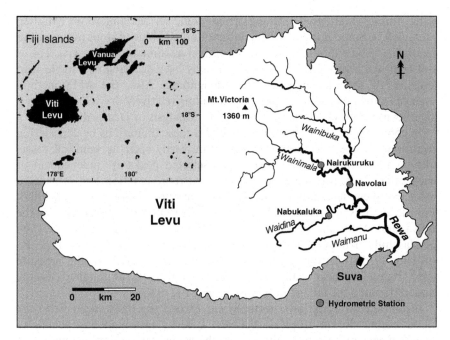

FIG. 9.6. The Rewa River on Viti Levu island in Fiji, showing its major tributaries and the sites of three long-term hydrometric stations operated by the Hydrology Division of the Fiji Public Works Department.

between the onset of intense rainfall and the rise of the rivers. On Grande Terre island in New Caledonia for instance, river discharge may increase 100-fold in less than an hour (ORE 2002).

The natural vegetation of many high islands across the South Pacific was originally forest, but extensive areas were cleaned and burned by early farmers. On the western drier sides of bigger islands like Viti Levu in Fiji this assisted the spread of savannah-like grasslands in the mid-Holocene. Further vegetation clearance and burning has been associated with the introduction of grazing animals, commercial cropping or the timber industry after the arrival of European colonists in the nineteenth century. These types of land-management practices and land-use changes exacerbate hydrological short-circuiting and enhance the size of cyclone-generated floods as a result.

9.2.2 Flood Analyses

With sufficient river discharge data from many tropical cyclones, it is possible to compare the measured peakflows with several types of extreme value probability distributions. For hydrological frequency modelling, the annual maximum series (AMS) and the partial duration series (PDS) are two approaches commonly employed. The AMS approach uses the largest annual value and is widely used, although it has some serious limitations (Madsen *et al.* 1997). Since it uses only the largest annual value, it does not consider secondary events that may exceed the annual maxima of other years. This is not helpful where several cyclone floods may have occurred in a single year, because all but the largest events are ignored. Also, the AMS method includes annual peaks from all years, including relatively dry years without a significant flood, which therefore biases the analysis of large values.

In comparison, the PDS approach avoids these limitations by considering peakflows above a certain (physical) threshold, such as bankfull stage, although it is important to ensure that selected peaks are independent of each other. It is also more flexible in flood representation and provides a more complete description of flood-generating processes (Lang *et al.* 1999). For the analysis of cyclone-induced floods, the PDS method was tested in the Rewa and Tontouta rivers in Fiji and New Caledonia, respectively by Kostaschuk *et al.* (2001, 2006). The recommended log-Pearson Type III probability distribution consistently provided the best fit to the measure data, based on mean daily flow records from 1970, although it usually underestimated the largest flood discharges (Fig. 9.7).

Instantaneous peakflows normally far exceed the daily mean flows for flood events during tropical cyclones. Clearly, instantaneous peakflows would therefore be more suitable for observing the maximum impacts of cyclones on flood generation. Unfortunately, there are few long-term hydrometric stations in the South Pacific islands adapted to measure very large events, and indeed there are many cases where stage-recording machinery has been completely destroyed by catastrophic floods, often leading to long gaps in the hydrological record afterwards.

The Liliane hydrometric station on the Tontouta River in New Caledonia is one site that was adapted to record large floods in the mid-1970s, so that instantaneous peakflows could be determined. In earlier years, flood marks were surveyed to calculate maximum discharges. The Tontouta River is located on the southwestern side of Grande Terre island. The basin is roughly 380 km^2 in area with steep slopes and rugged terrain in its upper reaches. Bankfull discharge at the hydrometric station is approximately 600 m^3 s^{-1}. Table 9.4 lists 37 instantaneous discharges for floods at Liliane between 1969 and 2003. Almost 65% of floods were caused by tropical cyclones, with TC Anne in 1988 producing a phenomenal discharge of nearly 4,600 m^3 s^{-1}. Expressed as specific discharge, this equivalent to a discharge of 12 m^3 of water every second per square kilometre of catchment area (12 m^3 s^{-1} km^{-2}).

9.2.3 Case Study – River Responses in Fiji to a Succession of Tropical Cyclones During the 1997 El Niño Event

In early 1997, as El Niño conditions developed across the Pacific region, five tropical cyclones named Freda, Gavin, Hina, Ian and June, traversed Fiji waters within a period of just 4 months. These storms were the first set

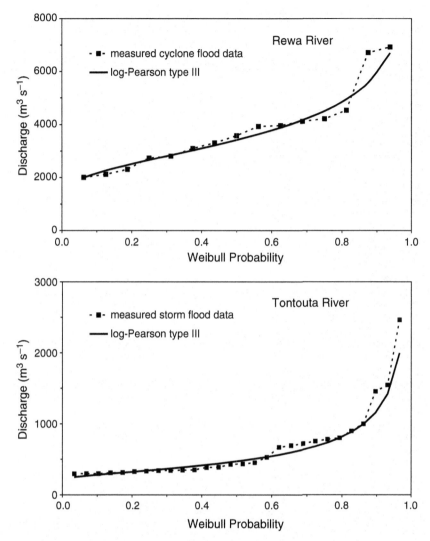

FIG. 9.7. Partial duration series for peakflows in the Rewa and Tontouta rivers in Fiji and New Caledonia (based on mean daily discharge data), generated mainly by tropical cyclones. From Kostaschuk *et al.* (2001, 2006).

of cyclones to severely affect the Fiji Islands since 1994. The majority of them brought fierce winds with intense rainfall, and induced big responses in many rivers.

The tracks of the five cyclones are shown in Fig. 9.8. TC Freda (20 January–1 February) formed to the northeast of the country and moved away to Vanuatu before curving back to Fiji with gale-force winds on 26 January.

TABLE 9.4. Maximum instantaneous discharge (Q_i) for the Tontouta River and tropical cyclones responsible for these peakflows, recorded between 1969 and 2003 at Liliane hydrometric station on Grande Terre island in New Caledonia.

Date and time	Q_i (m³ s⁻¹)	Tropical Cyclone (or unnamed tropical storm)
02/02/1969 04:30	551	Storm
02/01/1971 11:30	780	Rosie
06/02/1972 23:40	745	Daisy
03/06/1972 15:00	613	Ida
04/02/1974 08:45	1,744	Pam
08/03/1975 04:00	2,613	Alison
18/04/1975 01:30	3,18	Storm
17/01/1976 13:20	1,055	David
06/01/1978 23:00	514	Storm
12/02/1981 23:45	1,553	Cliff
07/03/1981 06:00	355	Freda
24/12/1981 22:00	2,420	Gyan
13/01/1988 00:00	4,583	Anne
15/11/1988 06:45	330	Storm
17/12/1988 14:54	417	Eseta
02/01/1989 22:48	852	Delilah
21/01/1989 16:30	899	Storm
12/02/1989 04:36	400	Harry
11/04/1989 05:18	716	Lili
23/01/1990 09:50	1,076	Storm
02/02/1990 05:44	582	Storm
25/02/1990 19:05	986	Storm
08/03/1990 05:05	478	Storm
17/02/1992 11:27	475	Daman
05/03/1992 04:57	928	Esau
10/03/1992 20:39	471	Fran
24/03/1992 06:21	637	Storm
08/04/1992 01:57	1,783	Storm
06/01/1994 11:34	686	Rewa
27/02/1994 13:10	1,010	Theodore
27/03/1996 19:36	3,356	Beti
08/01/1997 12:36	1,637	Drena
07/07/1997 12:06	593	Storm
25/01/1999 04:09	362	Dani
07/03/2002 00:16	909	Des
14/03/2003 11:25	3,166	Erica
16/07/2003 05:02	503	Storm

Source: Observatoire de la Ressource en Eau, New Caledonia.

An associated trough produced the wettest January on record for central and northwestern parts of Viti Levu island, which experienced 2–3 times the average January rainfall. TC Gavin (4–11 March) developed north of Fiji waters and west of Tuvalu, achieving cyclone status with storm-force winds at approximately 10°S 173°E, strengthening to hurricane intensity on 5 March. This was the worst storm to strike Fiji in 1997, with the eye passing close to

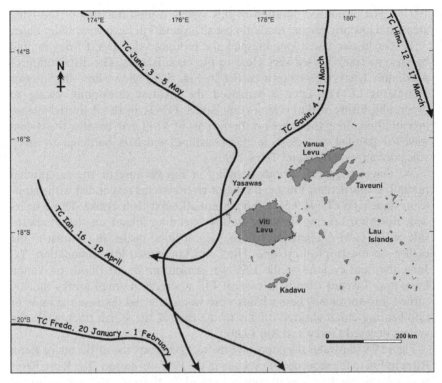

FIG. 9.8. Tracks of five tropical cyclones in the Fiji area during the first half of 1997, coinciding with the onset of a strong El Niño episode.

the north and west of Viti Levu on 7 March. TC Hina (12–17 March) brushed past the Lau group in eastern Fiji on its way to Tonga, but otherwise had little effect. TC Ian (16–19 April) skirted southern parts of the Fiji archipelago but left an active trough in its wake that caused widespread rain and some heavy falls. TC June (3–5 May) developed within Fiji waters and displayed erratic changes in direction and speed, remaining almost stationary over the tiny Yasawa islands on 6 May, before taking an unexpectedly sharp turn southwards. Although TC June was a relatively weak system, it bears the distinction of having been only the fourth cyclone since 1840 to threaten Fiji outside the normal hot season, which usually ends in April (Fiji Meteorological Service 1997a).

Available data on precipitation and river rises demonstrate the meteorological conditions and associated hydrological effects of these tropical cyclones. Weather records for Fiji's 22 synoptic climate stations reveal how the low-pressure trough associated with TC Freda gave heavy rain for 14 consecutive days from 19 January, setting new January monthly rainfall records in many parts of western Viti Levu (Fiji Meteorological Service

1997c). The prolonged duration of this system caused flooding of low-lying areas, and low bridges and roads almost all around Viti Levu were under water.

TC Gavin also had a long lifespan and produced the largest 1-day rainfall because its track passed very close to the main islands. The distribution of maximum 1-day falls (shown earlier in Fig. 5.28) shows how the interior mountains of Viti Levu experienced the greatest downpour, owing to orographic lifting of the cyclone's rain bands. TC Gavin also delivered intense precipitation along the entire northern coast of Viti Levu, because the system travelled parallel and close to this coastline, which is bordered by steep volcanic ranges (Ollier and Terry 1999).

TC Gavin triggered serious flooding in Fiji because of the exceptional rainfall. In particular, Viti Levu's major river systems responded with significant stage rises (Table 9.5), although not all burst their banks. Those rivers with northward-facing drainage basins extending inland on the windward side of TC Gavin's approach saw the largest flood peaks. In contrast to the earlier storms, tropical cyclones Hina and Ian caused little inundation. TC June, the final cyclone of the 1997 wet season, produced floods on Vanua Levu and Taveuni islands in eastern Fiji where even small creeks quickly turned into torrents. These islands were worst affected because the mass of rain-bearing cloud sheared off far to the east of the storm track while the system decayed (Terry and Raj 1999).

Figure 9.9 illustrates the impressive hydrological response of the major Rewa River in Fiji to the sequence of cyclones in the 1997 wet season. The Rewa River covers the largest drainage basin in the tropical South Pacific, 2918 km^2, approximately a third of the entire land area of Viti Levu island. The northern part of the watershed is underlaid mostly by relatively young basaltic rocks and the southern part by older plutonic rocks. Topography is hilly near the coast, but becomes steep and rugged in the upper basin, rising to 1,360 m at the Mount Victoria divide. Mount Victoria (called *Tomaniivi* in Fijian) is the tallest summit in the Fiji archipelago. Tributary watercourses are steep in their upper and middle sections (mean slope = 0.033) but become more gentle downstream in the alluvial reaches (mean slope = 0.002) where the gauging stations are located. Dense rainforest vegetates the highlands, whereas subsistence cultivation and cattle pasturing are important in valley bottoms.

TABLE 9.5. Peak flood stages in four of Fiji's principal rivers during Tropical Cyclone Gavin on 8 March 1997.

River system	Basin area (km^2)	Peak river stage (m)
Nadi River	490	6.66
Ba River	930	6.30
Rewa River	2,918	4.52
Sigatoka River	1,450	3.44

All river levels are measured as heights above fixed benchmarks at established gauging stations. Source: Hydrology Division, Fiji Public Works Department.

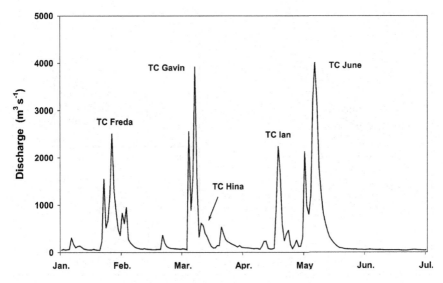

FIG. 9.9. Peak daily flows in the Rewa River at the Navolau hydrometric station, produced by a succession of tropical cyclones during the El Niño event of 1997. Bankfull discharge is approximately 2,000 m^3 s^{-1}.

The Rewa has four major tributaries: the Waimanu, Waidina, Wainimala and Wainibuka rivers. These drain southern coastal, southern interior, interior highland and northeastern portions of Viti Levu, respectively. The hydrometric station at Navolau on the main Rewa channel, several kilometres below the confluence of the Wainimala and Wainibuka tributaries, records river discharge from an area of 1,960 km^2 (Fig. 9.6). Bankfull discharge is estimated at 2,000 m^3 s^{-1}. At this site there is a well-recorded history of major floods brought by tropical cyclones. Flooding has occurred 15 times during cyclones between 1970 and 2000, or 1 cyclone-related flood every 2 years on average.

Figure 9.9 illustrates three points. First, 4 out of 5 tropical cyclones in the 1997 El Niño episode produced overbank floods. For 1997 this gives a cyclone-flood frequency 8 times greater than average. Second, hydrograph peaks for the cyclone events are generally higher than maximum discharges produced by other types of tropical storms during the same wet season (the unlabelled peaks on the hydrograph). Third, tropical cyclone floods caused widespread inundation because they were significantly larger than bankfull discharge. For instance, cyclones Gavin and June generated peakflows that were twice the channel capacity at Navolau.

The Nakauvadra and Teidamu rivers are minor fluvial systems compared to the Rewa River, but this makes them useful for investigating the behaviour of small rivers receiving cyclone-intensity precipitation. Both rivers drain catchments on the northern side of Viti Levu island (Fig. 9.10). Above its gauging station, the Nakauvadra River has a watershed covering 38 km^2 and

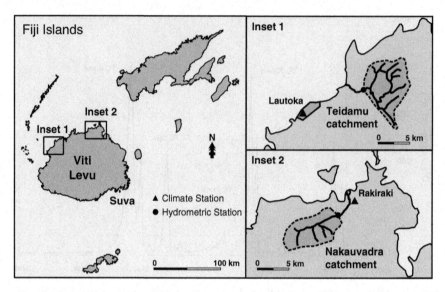

FIG. 9.10. Location of the Teidamu and Nakauvadra rivers in northern Viti Levu island, Fiji.

flows from west to east, reaching the coast by Rakiraki town. The Nakauvadra mountains run parallel to the coast about 7 km inland and rises to a maximum elevation of 866 m. This range of andesitic rocks is the eroded rim of the large Rakiraki volcano, which forms angular terrain along the southern watershed boundary. Drainage concentration at the base of the foothills supports dense rainforest. Less rugged landscapes nearer the coast are cultivated with sugarcane and subsistence crops. The catchment of the Teidamu River above the gauging station drains 56 km^2 of northwestern Viti Levu, close to the city of Lautoka. The fertile lower parts of the catchment are farmed for sugarcane, while upland slopes are planted with commercial pine forest (Terry and Raj 2002). The inland watershed is marked by a north–south ridge climbing to 480 m, which is the northerly extension of the Mount Evans range farther south.

Several observations are possible from the pair of hydrographs in Fig. 9.11. First, the Nakauvadra and Teidamu rivers both display extremely flashy behaviour, reflecting their short and steep watercourses, and the small size yet rugged terrain of their catchments. Second, the magnitudes of peak daily flows vary considerably between individual storm events. Cyclones Freda and Gavin were long-lived systems, and therefore produced outsize responses in these small rivers, whereas TCs Hina, Ian and June with short lifespans had far less effect. Third, double or multiple hydrograph peaks may be produced during the life of a single cyclone, such as TC Freda. Multiple peaks show that the most intense cyclonic rain falls sporadically, probably associated with individual cells of very heavy precipitation embedded in the mass of

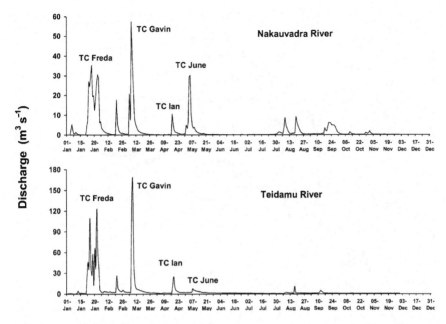

FIG. 9.11. Peak daily flows in the Nakauvadra and Teidamu rivers in Fiji, generated by the sequence of tropical cyclones during the early phase of the 1997 El Niño event.

rain-bearing cloud. In consequence, river floods may recede for some time during quiescent periods, only to rise dangerously again later, even as a cyclone enters its decay stage.

9.2.4 Case Study – Exceptional River Flooding on Vanua Levu Island Caused by Tropical Cyclone Ami in January 2003

Tropical Cyclone Ami was the third cyclone to form in the Fiji area during the wet season of 2002–2003. A tropical depression was first identified at about 9 a.m. Fiji Standard Time (FST) on 10 January 2003. The system was embedded in an active low-pressure trough, 386 km east of Funafuti atoll in Tuvalu. Its development was affected by diurnal variations and strong vertical shear, but on 12 January the depression underwent rapid maturation and was subsequently named TC Ami at about 12 midday FST, when it attained gale strength near Niulakita island in southern Tuvalu (Fig. 9.12).

Initially the cyclone followed a southwest track at about 15 km h[-1], but gradually curved southwards as it approached the island of Rotuma. Once named, TC Ami deepened rapidly to storm intensity before midnight on 13 January, and then to hurricane intensity during 14 January. As the cyclone accelerated along its southerly path, the radius of its destructive winds increased.

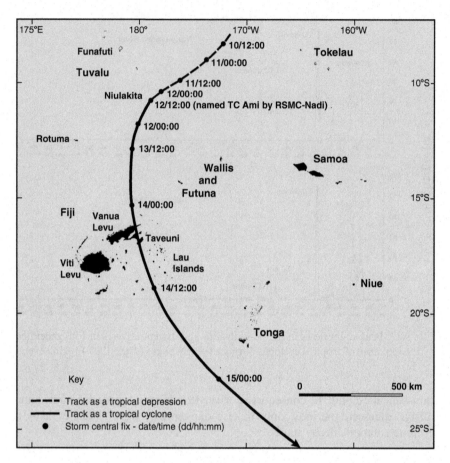

FIG. 9.12. Track of Tropical Cyclone Ami through the tropical South Pacific from 10 to 15 January 2003.

The eye of the storm made landfall near Dogotuki on the northeast peninsula of Vanua Levu island shortly after 3 a.m. on 14 January, bringing ferocious hurricane-force winds to much of Fiji's northern division. The eye passed over the western tip of Taveuni island after 5 a.m. on 14 January. The cyclone then moved quickly through the Lau islands (Fig. 9.13), veering to the southeast as it did so. At 12 noon on 14 January, TC Ami attained its peak intensity, when its centre was located 97 km south–southwest of Lakeba island. Sustained winds of 200 km h^{-1} were reported, with momentary gusts of 230 km h^{-1} (NIWA 2003). Thereafter TC Ami accelerated further, turned south–southeast and travelled at 22 km h^{-1} as it departed Fiji waters, all the while maintaining hurricane intensity.

Across Fiji's northern and eastern divisions the destruction caused by TC Ami was extensive and communications were completely cut off for several

FIG. 9.13. Visible satellite image of Tropical Cyclone Ami, illustrating the organisation of the cloud bands at approximately midday on 14 January 2003 (Fiji Standard Time). The eye of the storm is passing through the Lau group of islands in eastern Fiji on a south–southeasterly track. Base image courtesy of NOAA.

days. The confirmed number of fatalities was 17 people. Fiji's National Disaster Management Centre declared Vanua Levu island a zone of natural catastrophe. One-day rainfall figures for climate stations on Vanua Levu are given in Table 9.6. The data indicate that large-scale rainfalls were widespread. Of the 18 stations listed, 16 recorded more than 100 mm of rain in

TABLE 9.6. Maximum 1-day rainfall (9 a.m.–9 a.m.) delivered by Tropical Cyclone Ami in January 2003, measured at climate stations on Vanua Levu and Taveuni islands.

Climate station location	Island	Reference no. in Fig. 9.14	Rainfall (mm)	Date in January 2003
Vatuwiri	Taveuni	1	311	13
Seaqaqa Forestry Station	Vanua Levu	2	270	13
Labasa Airport	Vanua Levu	3	245	15
Wailevu	Vanua Levu	4	214	14
Vunimoli	Vanua Levu	5	200	15
Labasa Sugar Mill	Vanua Levu	6	194	14
Naravuka	Vanua Levu	7	192	13
Waiqele	Vanua Levu	8	192	13
Nagigi	Vanua Levu	9	175	13
Tutu	Taveuni	10	167	13
Batiri Citrus Farm	Vanua Levu	11	162	13
Natua	Vanua Levu	12	145	14
Wainikoro	Vanua Levu	13	132	14
Kurukuru	Vanua Levu	14	124	15
Seaqaqa Agriculture	Vanua Levu	15	109	13
Seaqaqa Sub-Station	Vanua Levu	16	109	14
Nabouwalu Port	Vanua Levu	17	79	13
Rokosalese	Vanua Levu	18	63	13

Source: Fiji Meteorological Service.

24 h. Of these, five stations received more than 200 mm. The majority of climate stations experienced their heaviest downpours on 13 January. Maximum recorded precipitation was 311 mm at the coastal site of Vatuwiri on Taveuni island. Since TC Ami approached Fiji from the north, Vatuwiri lay on the windward coast beneath the tallest mountains in the region (1,241 m), so the large rainfall recorded there reflects orographic enhancement by the volcanic relief (Terry *et al.* 2004).

Vanua Levu's geology is made up of volcanic rock types, mainly in the form of lava flows, pyroclastics, breccias and conglomerates. Its geomorphology is dominated by a chain of volcanic mountains aligned in a southwest to northeast orientation, forming a central highland spine along the island and giving it a mountainous profile. The three tallest peaks are located towards the centre of the chain, south of Labasa town. They are Delaikoro (941 m), Koroalau (1,032 m) and Dikeva (957 m). Most river networks drain northwest or southeast, controlled by the linear arrangement of the volcanoes (Fig. 9.14). River basins are separated by sharp, irregular interfluves. Highland slopes are steep, frequently above 35°. Headwater channels are extremely bouldery. Lower watershed areas have undulating terrain and flat areas of floodplains and alluvial terraces. The mountains are covered with rainforest, whereas the coastal lowlands have a dense patchwork of commercial sugarcane fields, interspersed with savannah-type grasslands.

Flooding is especially perilous for the coastal hinterlands around Labasa owing to several physiographic influences. First, the north coast of Vanua

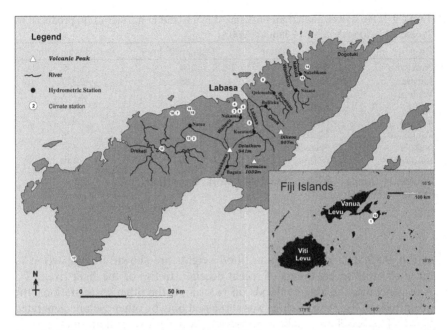

FIG. 9.14. Synoptic climate stations, principal river systems and their hydrometric stations on Vanua Levu island in northern Fiji.

Levu is highly indented with many embayments. This configuration increases the possibility that storm surge inundation will combine with river floods (Terry and Raj 1999). Second, three large rivers, the Wailevu, Labasa and Qawa rivers, all empty into estuaries located in the same area. Third, these rivers all rise in the highest mountains on the island and fourth, their basins face northwest in the direction from which many cyclones arrive. Orographic effects therefore often produce enormous rainfalls that are promptly converted into large hydrological responses because of the physical characteristics of landscape. Fifth, replacement of natural grasslands on lower hillslopes by sugarcane plantations exacerbates runoff and erosion (Morrison 1981).

Due to the torrential rainfall dumped by Tropical Cyclone Ami in January 2003, some of the worst-ever flooding occurred in many rivers on Vanua Levu island. Drowning was the main cause of death for the 17 fatalities. Peak discharges of the eight main rivers on Vanua Levu are presented in Table 9.7, to indicate the magnitude of the floods produced by TC Ami. Maximum flows were calculated from the heights and slope of mud lines, surveyed by the Hydrology Division of the Fiji Public Works Department at their long-term hydrometric stations, shortly after the floodwaters receded. Locations of the rivers and their gauging stations are given in Fig. 9.14.

To provide some idea of how these peak river discharges corresponded to flood height, three river cross-sections at different gauging stations are drawn

TABLE 9.7. Peak river discharges on Vanua Levu island in northern Fiji, produced by Tropical Cyclone Ami on 14 January 2003.

River	Hydrometric station	Basin area[a] (km²)	Peak discharge (m³ s⁻¹)
Nasekawa	Bagata	104	6,139
Labasa	Korotari	86	2,377
Wailevu	Nakama	77	2,118
Qawa	Bulileka	38	1,802
Dreketi	Natua	128	996
Wainikoro	Nasasa	45	676
Nakula	Nakelikoso	16	559
Bucaisau	Qelemumu	80	447

Source: Hydrology Division, Fiji Public Works Department.
[a]Above hydrometric station.

in Fig. 9.15. TC Ami maximum flood heights are shown in comparison to those of other severe events of recent decades. In five of the eight rivers, TC Ami generated the biggest floods on record. At the other three stations, the magnitude of TC Ami's deluge was surpassed only by other cyclone-generated floods. The Nasekawa, Labasa, Qawa and Wailevu rivers all drain the mountainous terrain in the middle of Vanua Levu. The Nasekawa River produced a huge peak discharge of more than 6,100 m³ s⁻¹, equivalent to almost 60 m³ of water per second for each square kilometre of the catchment (60 m³ s⁻¹ km⁻²). This mighty torrent tore down the main bridge on the south coast highway at Bagata village.

On the north coast, the Labasa and Qawa rivers both drain into the same sheltered bay near Labasa town, and the estuary of the Wailevu River lies only 4 km farther west along the coast near Labasa airport. The climax flows in these rivers were all record-breaking, respectively 2,377, 1,802 and 2,118 m³ s⁻¹. The simultaneous discharge of these spectacular volumes of water from all three rivers onto the same coastal plain coincided with a strong storm surge that was felt along the entire north coast of Vanua Levu. This caused inundation to depths of 3–4 m above the floodplain over a wide area around Labasa (Fig. 9.16). Loss of life and unprecedented despoliation of farms and homes across many rural communities was inflicted. Extensive ruin of infrastructure and property was suffered in Labasa town, the primary urban and commercial centre for Vanua Levu and the whole northern division of Fiji.

9.2.5 Flood Hazard Mitigation

The historical record of tropical cyclones in the South Pacific indicates that storms rarely develop close to the main island archipelagoes lying south of 10°S. Instead, nascent tropical disturbances tend to form farther north and then approach on southerly tracks over 2 or 3 days as they mature. This crucial delay should prove invaluable for alerting vulnerable populations

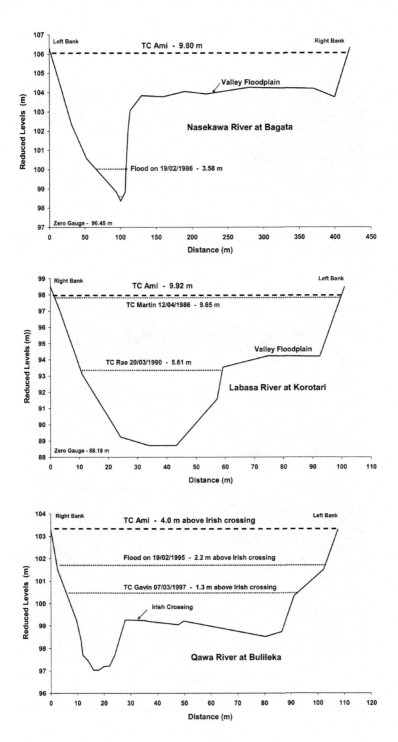

FIG. 9.15. Maximum flood heights in several rivers on Vanua Levu island in Fiji, generated by Tropical Cyclone Ami in January 2003. Comparison is made with other major cyclone-induced floods. Source: Hydrology Division of the Fiji Public Works Department.

FIG. 9.16. Severe inundation of Labasa town on the northern coast of Vanua Levu island in Fiji, resulting from the combined effects of inundation by the sea (storm surge) and flooding of the Labasa and Qawa rivers. Photos courtesy of the Fiji Meteorological Service.

inhabiting low-lying coastal areas and valley bottoms, in advance of expected floods. Unfortunately, dispersed rural communities are often insufficiently prepared to cope with flood hazards. All too commonly, either people ignore warnings issued by national meteorological offices because they are unwilling to leave their homes, farms and property, or others who are better educated in the dangers do not have adequate mobility to escape to higher ground. As inundation waters rise, many families become trapped in their houses, panic and then try to wade to dry ground sometimes with fatal consequences.

In the future, the magnitude of hydrological hazards for the high Pacific islands may increase if ocean warming leads to more intense tropical cyclones (Kostaschuk *et al.* 2001). Based on this scenario, improved disaster-reduction programmes are needed in all island nations to avoid escalating loss of life. A survey by the World Bank (2000) suggested a range of options for moderating the flood impact of cyclones. These are grouped according to the following categories.

9.2.5.1 Flood Control

There are several engineering measures for controlling river floods, including both structural and 'soft' techniques. River diversion channels, retarding basins, cut-off channels and retention dams are structural controls, whereas raising embankments, river channel widening and riverbed excavation are soft engineering measures (JICA 1997). Although costly to build, retention dams may be one of the most beneficial options for Pacific Islands where suitable topography exists for their construction, because dams can be used for water-resource development as well as for controlling floodwaters (Terry 2002).

9.2.5.2 Catchment Management

Improving catchment management involves the combination of several activities, such as reforestation, soil conservation, regulating land development and protecting natural wetlands. These measures help to improve the natural water-retention function of drainage basins and thereby maintain existing flow capacities of rivers by avoiding excessive silting up of channels.

9.2.5.3 Mitigation

It is possible to reduce the potential damage caused by floods by restricting the urbanisation and settlement of low-lying areas and by promoting the use of flood-proof house design where necessary. The resilience of social infrastructure can also be increased through community-education programmes to raise awareness of tropical cyclone characteristics and behaviour. This needs to take place in conjunction with the introduction of better communication systems within relevant government institutions, so that the public may be more effectively warned about impending flood hazards, both before and during tropical cyclone events.

Chapter 10
Fluvial Geomorphology

10.1 Channel Adjustments

River channel morphology and pattern must adjust to major flood events on tropical Pacific islands, generated by extreme rainfall during tropical cyclones. Alluvial rivers often exhibit large changes in their channels, especially in terms of geometry and position, because they are sensitive to the erosive power of huge river discharges during cyclones. Riverbanks are undercut and collapse (Fig. 10.1), meander bends are cut off and abandoned and riverbeds scour and fill. Yet considering the importance of cyclone-induced floods and related river channel instability for human occupancy and activities on floodplains, it is a pity that little quantitative information exists on the nature and rates of river channel change in the South Pacific.

The Wainimala River, in the large Rewa basin on Viti Levu island in Fiji (Fig. 10.2), is one example where rapid movement in channel position has been documented, primarily because shifting channels have disrupted farming activities and led to consequent disputes among the local population. This is because rivers and streams across the Pacific islands are frequently used to demarcate the boundaries of tribal land between adjacent clans, and so changing river patterns can have important repercussions for land ownership.

Over the 20-year hydrological record 1978–1997 for the Wainimala River, bankfull and overbank floods measured at the Nairukuruku hydrometric station have an average return period of 2 years. Nine floods were caused by tropical cyclones and one other by a wet-season depression. The hydrological response to Tropical Cyclone Gavin in March 1997 is shown in Fig. 10.3 as an example. The Wainimala has a basin area of 810 km^2 above the hydrometric station, and bankfull discharge at the gauging site is 1,200 $m^3 s^{-1}$. It is seen that river flow rose rapidly from 50 to 6,300 $m^3 s^{-1}$, producing a peakflow that exceeds five times the bankfull capacity.

According to oral accounts by the local people, flood events of this magnitude are highly erosive. Associated geomorphic responses in the Wainimala River have been riverbank retreat, channel migration across the floodplain, and large-scale adjustments in channel-planform geometry.

FIG. 10.1. Extensive riverbank slumping along the Wainimala River in the Rewa basin, eastern Viti Levu island in Fiji. Bank collapse was caused by the huge discharge during Tropical Cyclone Gavin in early March 1997. Note the woman fishing in the centre of the photograph for scale. Photo by Ray Kostaschuk.

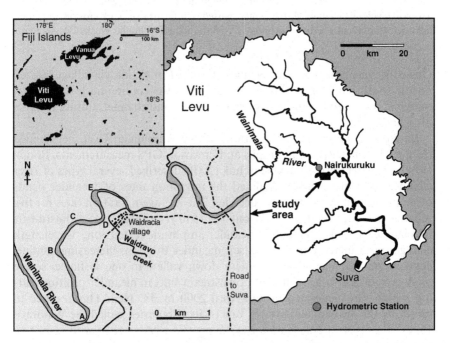

FIG. 10.2. The study area for channel migration on the Wainimala River in Fiji. The inset map shows the position of the Wainimala channel and its Waidravo tributary in 1994. The main Wainimala has subsequently migrated downstream and now occupies the section of Waidravo creek adjacent to Waidracia village (see Fig. 10.4).

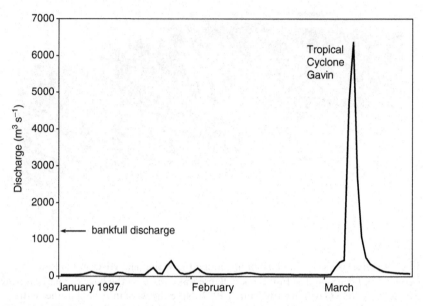

FIG. 10.3. Hydrograph of the Wainimala River at Nairukuruku hydrometric station in the early months of 1997. The hydrograph illustrates the extreme peakflow generated by Tropical Cyclone Gavin in March, which caused significant riverbank erosion and channel migration. Data source: Hydrology Division of the Fiji Public Works Department.

One consequence is that Ratu Nemani, the chief of Waidracia village located on the south bank of the river, has lamented the continuing loss of river-marginal farmland because of riverbank retreat and channel shifting towards his village.

Based on surveys of past positions of the Wainimala River, determined from historical aerial photographs and precision GPS measurements, Rodda (1990) and later Terry and Kostaschuk (2001) described several types of river channel adjustments and calculated the migration rates of meander bends over approximately the last five decades. The average migration rates for five separate meander bends ranged from 5 to 15 m yr^{-1} (Table 10.1). Adjustments in channel pattern included meander-amplitude extension, wavelength change (both increase and decrease), meander rotation, increasing channel convolution and meander movement down valley. In one instance, severe bank erosion led to a significant downstream shift in meander position during a high flow generated in mid-April 2000 by TC Neil. This resulted in piracy by the Wainimala River of one of its tributaries, called the Waidravo Creek (Fig. 10.4). *Channel piracy* is the process where a river with a shifting channel eventually erodes into, and then occupies, the channel of another river. The consequence for the Waidracia people was that they lost ownership of a key area of farmland adjacent to their village, where they previously grew food crops and grazed their cattle.

TABLE 10.1. Migration and geometric change in five individual meander bends in the Wainimala River, Fiji, measured over the period 1954–1994.

Meander bend	Total movement[a] (m)	Channel migration rate (m yr⁻¹)	Meander planform change
A	250	6	Amplitude ↑
B	200	5	Wavelength ↑
C	600	15	Amplitude ↑ and wavelength ↓
D	550	14	Amplitude ↑ and wavelength ↓
E	300	8	Axis rotation and sinuosity ↑

The position of the channel in 1994 is shown in Fig. 10.2.
[a]Measurements to the nearest 50 m.

FIG. 10.4. Low-level aerial view in July 2001 of the Wainimala River adjacent to Waidracia village, in the Rewa basin, Fiji. Prior to early 2000, the river occupied the big meander loop in the centre of the photograph. A minor tributary channel originally flowed from right to left in front of the village, joining the main river farther downstream. Down-valley migration of the meander bend occurred during a high flow caused by rainfall associated with Tropical Cyclone Neil in mid-April 2000. This led to piracy of the tributary channel and abandonment of the old meander in preference for the present straight course. The area of floodplain farmland enclosed by the cut-off loop is now effectively 'lost' from Waidracia village ownership. Photo by Ray Kostaschuk.

10.2 River Sediment Transport

Those people who live in tropical islands with rivers and streams are familiar with one important feature of hydrological behaviour during tropical cyclones – when rivers begin to rise soon after heavy rainfall starts, the water becomes more turbid at the same time. This is because during the rising stage of a storm hydrograph, the sediment concentration carried by a river generally increases. In addition, powerful cyclone-induced floods are among the few occasions when coarse bedload sediments are set in motion and transported downstream (Fig. 10.5). Exceptional amounts of suspended sediment and bedload sediment transport reflects the ability of cyclone floods to erode and reshape features in the fluvial landscape. Perhaps not surprisingly, there exists only a limited number of measurements of sediment transport during tropical cyclones in South Pacific rivers, primarily because of the logistical difficulties in trying to sample overbank flows and within-channel sediment movement during perilous flood conditions.

Even where sediment sampling has been successful, relations between flow and suspended sediment concentrations during tropical cyclone events can sometimes be very difficult to interpret for a variety of reasons. Often the sampling efficiency of apparatus used is uncertain, so at least some of the range in suspended sediment values could be due to variability in sampling precision. Indeed, many types of rising-stage bottle apparatus, for example,

FIG. 10.5. Coarse gravels and boulders in Somosomo creek on Taveuni island in northern Fiji. This material was excavated from the riverbanks by floodwaters and transported downstream as bedload during Tropical Cyclone June in May 1997.

sample only the surface of the flow. Yet the concentration of suspended particles varies considerably within a river's cross section, depending on the erodibility of the riverbed and floodplain, the shear stress of the flow and the height above the channel bed. Furthermore, at any individual study site, the ratio of the supply of fine sediment to the total suspended load of the river will fluctuate over the duration of a flood event, depending on bank failure, erosion patterns in the watershed and many other factors. This can give relationships between suspended sediments and river discharge that are *hysteretic* in form. This means that the peak in suspended sediments may either precede or lag the peak in water discharge.

Notwithstanding the uncertainties mentioned above, investigations in the Rewa River in Fiji have been able to provide some information on suspended sediment transport during tropical cyclones. As part of an environmental assessment of the Rewa basin by a private consulting company (NSR Environmental Consultants 1994), rising-stage water samplers were installed on 10 October 1992 and operated until 6 January 1994. During the period of the project, two tropical cyclones struck. These were TC Joni and TC Kina (Fig. 10.6), the latter being one of the fiercest storms to afflict Fiji in recent decades.

Both the tropical cyclones generated impressive overbank floods in the Rewa River, recorded at Navolau hydrometric station (Fig. 10.7). TC Joni's precipitation began slowly on 8 December 1992, so river levels did not rise until rainfall intensified on 10 December. Most of TC Joni's rain fell on 11 December, the same day that river flow peaked. Rainfall ended on 12 December and river flow declined rapidly. TC Kina's rainfall commenced on 29 December 1992 and continued until 1 January 1993, producing a small flow peak on 31 December. The heaviest rain was delivered on 3 January and resulted in an enormous discharge event on the same day. The storm hydrograph then fell sharply.

A total of 12 rising-stage water samples were collected from TC Joni and 13 from TC Kina. Laboratory analysis of the particle grain sizes showed that the samples comprised both fine materials, sometimes called the wash load, and suspended riverbed sands. The sand fractions within the samples reflect the large shear stresses generated in the river flow during the floods (Kostaschuk *et al.* 2003).

A graph of temporal fluctuations in suspended sediment concentration during the rising stages of the two floods is provided in Fig. 10.8 . Values of sediment concentrations were low as the stage began to rise (S^1 = 0.92 m) during the TC Joni event on 10 December 1992. These increased as the hydrograph steepened (S = 1.42 m), but then fell consistently as river levels rose during the early hours of 11 December (S = 5.05 m). The sediment

---

[1] S refers to the height of the river surface (called the *stage*) above a fixed benchmark.

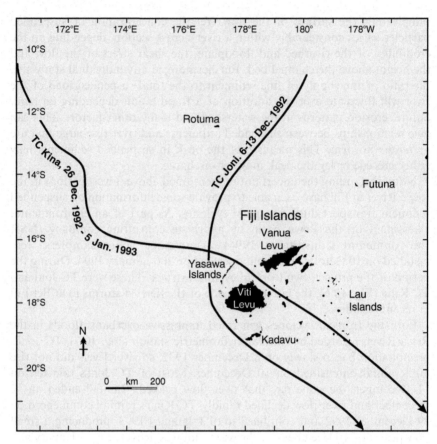

F<small>IG</small>. 10.6. Tracks of tropical cyclones Kina and Joni across the Fiji Islands in 1992–1993.

concentrations in two of the three samples from high flood stages (S = 8.82–9.82 m) reached over 80 g L^{-1}.

For TC Kina, the first, little flow peak between 29 December 1992 and 1 January 1993 (S = 1.42–6.04 m) gave the greatest values of sediment concentrations, with a peak value of over 950 g L^{-1} early in the flood. Later on during the main period of the flood (S = 10.82–11.82 m), the samples had much lower sediment concentrations. This probably reflects exhaustion of sediment supply from channel and catchment sources. The suspended sediment concentrations measured in the Rewa River during these tropical cyclones are some of the highest ever recorded during river floods, in both tropical and temperate regimes (Kostaschuk *et al.* 2003).

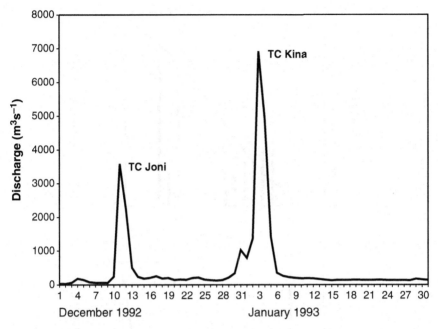

FIG. 10.7. Hydrograph of mean daily flows in the Rewa River, Fiji, during tropical cyclones Joni and Kina 1992–1993, measured at Navolau hydrometric station. Bankfull discharge is approximately 2,000 m^3 s^{-1}. Data source: Hydrology Division of the Fiji Public Works Department.

10.3 Valley Aggradation

10.3.1 Features and Rates

River floodplains are valuable components of the physical environment on mountainous islands in the South Pacific, where the availability of flat land is otherwise limited. They are favoured for the sites of towns and villages, and their alluvial soils are generally fertile and prized for farming. Floodplains are constructional features in the landscape and one of the chief processes by which they grow is the vertical accumulation of fine deposits laid down during successive overbank flood events. The upward rate of floodplain growth depends on the frequency of flooding and the suspended sediment concentration in the floodwaters. Tropical cyclones are therefore very significant morphological agents for floodplains, because cyclone-generated floods deposit enormous quantities of sediment over wide areas. Figure 10.9 illustrates the point, where thick layers of sand and mud were dumped on the valley floor of the Labasa River on Vanua Levu island in Fiji during the flood produced by TC Ami in January 2003.

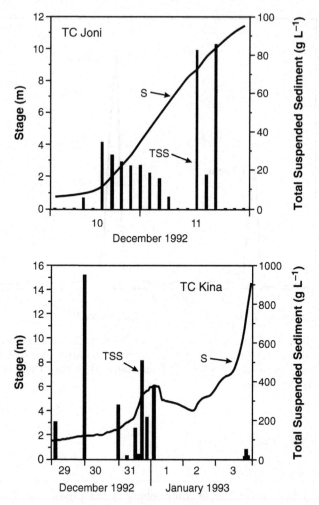

FIG. 10.8. Total suspended sediment (TSS) concentrations and equivalent river stage height (*S*) at Navolau hydrometric station. The data are plotted through time for the rise of the Rewa River during tropical cyclones Joni and Kina in the 1992–1993 wet season. Data source: NSR Environmental Consultants (1994) and Hydrology Division of the Fiji Public Works Department.

Tropical Cyclone Brenda generated a very big flood on 19 January 1968, in the Dumbéa River of southeast Grande Terre island in New Caledonia. Post-event fieldwork carried out by Baltzer (1972) confirmed that sediments are not left behind by floodwaters in a random fashion, but instead form into identifiable depositional structures with distinctive spatial patterns. Baltzer showed that the types of structures and their locations on the Dumbéa floodplain were related to differences in kinetic energy within the

FIG. 10.9. Deposition of large quantities of sand and mud on the floodplain of the Labasa River near Korotari on Vanua Levu island in Fiji, by the flood in January 2003 produced by Tropical Cyclone Ami. Photo by Pradeep Chand.

floodwaters. Kinetic energy was calculated from flood marks and measurements of floodwater-surface slope.

In the highest-energy areas, there was evidence of floodplain scouring and adjacent deposition of sand, pebbles and cobbles. These materials were arranged in a variety of ripples, small dunes and bar-like features. These structures covered small surfaces, but the deposits were in thick layers. In contrast, where floodwater currents spread out in a fan-like configuration, the inundation was less vigorous, so dunes and bars were absent. Instead, layers of finer sediments were more widely deposited. For floodwaters with average and low-energy currents, 1–10 cm thicknesses of fine sands and 1–3 mm layers of silts and clays were laid down, respectively over extensive portions of the floodplain surface.

If we are prepared to accept that tropical cyclone floods are a primary mechanism for overbank sediment deposition and floodplain growth, then it is illuminating to determine the long-term accretion rates associated with many floods. A small but growing body of evidence suggests that fluvial-sedimentation rates on tropical Pacific islands are some of the highest recorded globally. This appears to reflect both the comparatively high frequency of cyclone-induced floods and the elevated concentrations of suspended sediments during these events, as described in Section 10.2.

The rate of alluvial sedimentation over the last 40–50 years in actively aggrading valleys in Fiji, Samoa and Vanuatu has been recently investigated (Terry *et al*. 2002b, 2005). The ^{137}Cs method was employed, which is simple to grasp and has been used with success to establish sedimentation rates in a range of different environmental settings. In brief, floodplain alluvium is sampled in continuous increments down from the surface, and each incremental sample has its content of ^{137}Cs measured in a spectrometer. Cæsium-137 (^{137}Cs) is a radioactive element that does not occur naturally, but was first produced by the atmospheric nuclear weapons testing of the 1950s and 1960s. The year 1964 was the year of maximum ^{137}Cs fallout from the atmosphere across the tropical South Pacific. This means that the specific horizon in a vertical profile with the maximum ^{137}Cs content is normally taken to represent the 1964 surface of the palaeofloodplain. All the sediment lying above this horizon has been deposited since that time. Dividing the thickness of overlying sediment by the number of years from 1964 until present provides an average rate for floodplain build-up at the measurement site, in units of millimetres or centimetres per year (mm yr^{-1} or cm yr^{-1}).

Average floodplain-sedimentation rates of 1.0–5.8 cm yr^{-1} were determined from ^{137}Cs profiles, measured in the valleys of four rivers across the South Pacific (Terry *et al.* 2002b, 2005). These are the Falefa River on Upolu island in Samoa, the Jourdain River on Espiritu Santo island in Vanuatu, and the Wainimala River and Labasa River in Fiji on Viti Levu and Vanua Levu islands, respectively. Figures 10.10 and 10.11 illustrate the location of floodplain sampling in the Falefa valley in Samoa and the vertical distribution of ^{137}Cs content within the alluvium. Table 10.2 compares aggradation rates in river basins in wet-tropical regimes, and indicates that rates are comparatively high in the South Pacific islands.

10.3.2 Case Study – Catastrophic Valley Aggradation on Guadalcanal Island Caused by Tropical Cyclone Namu in May 1986

Guadalcanal island in Solomon Islands is mostly steep and rugged. The southern half of the island is a mountainous zone rising to over 2,300 m with a northwest-to-southeast trending spine. The mountains are flanked on the northern side by foothills that form an intermediate zone of intensely dissected plateaux, hills and rolling ridges (Hackman 1980). Numerous rivers transect this zone, draining generally northwards from the mountains (Fig. 10.12). The northern region of Guadalcanal is known as the Guadalcanal Plains and is an alluvial zone. From the Lungga River in the west to the Matepono River in the east, the plains are over 50 km wide. The topography here has minimal relief, except for some fluvial dissection in places. The rivers meander extensively across the plains, so swamps and poorly drained areas are common.

Tropical Cyclone Namu from 17 to 20 May 1986 resulted in the worst catastrophe to afflict parts of Solomon Islands in recent decades (track shown in Fig. 8.5). Unofficial tallies put the number of deaths at one hundred, but many more probably perished. On the north coast of Guadalcanal 350 mm of rain fell. The climate station at Gold Ridge, at a relatively low

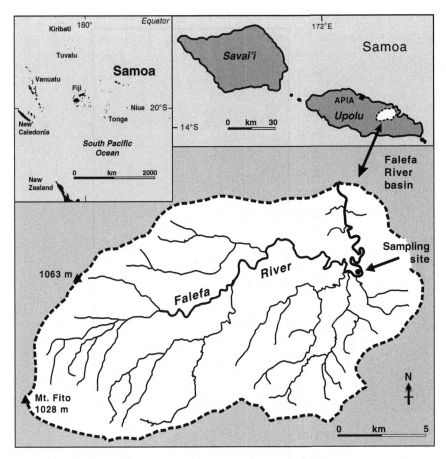

FIG. 10.10. Drainage map of the Falefa River basin on Upolu island in Samoa, showing the site of floodplain accretion measurements by Terry *et al.* (2005).

elevation of 290 m, recorded 874 mm of precipitation over the duration of the cyclone. Farther up in the mountains, moisture receipt is assumed to have been much greater still. TC Namu's torrential rain caused swollen rivers on the Guadalcanal Plains to burst their banks on 18–19 May. Extensive areas were awash with a metre of water, ruining 90% of food gardens and sweeping away homes and half the farm livestock (National Disaster Council 1986). Huge river discharges smashed bridges on the Lungga, Ngalimbiu and Mbalisuna rivers (Fig. 10.13). Retreating floodwaters left behind 30–50 cm thicknesses of silt over the plains (Table 10.3). Evacuation of thousands of people led eventually to permanent relocation by about 5% of the population away from the Guadalcanal Plains.

A survey of the geomorphic effects of TC Namu by Trustrum *et al.* (1989) provides an illustration of the extreme magnitude of river-channel

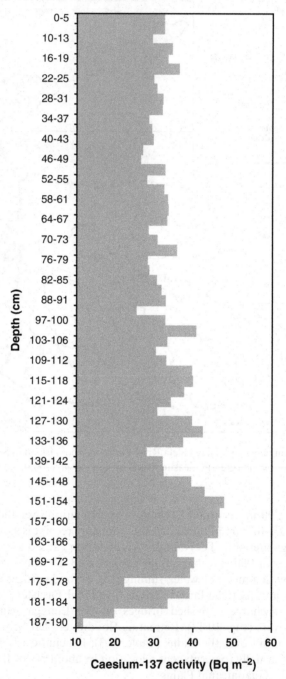

FIG. 10.11. Vertical profile of ^{137}Cs activity in floodplain alluvium adjacent to the Falefa River, on Upolu island in Samoa. From Terry *et al.* (2005).

TABLE 10.2. Recent historical rates of overbank sedimentation in various river basins in tropical climatic regimes.

River system	Location	Climate	Depositional site	Accretion rate (cm yr⁻¹)	Approximate timescale (or specific event)	Authors
Long-term measurements						
Krishna and Cauvery	India	Tropical monsoon	Delta	0.35–1.1	100 years	Vaithiyanathan *et al.* (1988)
Brahmaputra-Jamuna	Bangladesh	Tropical monsoon	Floodplains	0.67–1.15	50 years	Allison *et al.* (1998)
Brahmaputra	Bangladesh	Tropical monsoon	Floodplain	0.16	50–100 years	Goodbred and Kuehl (1998)
Hanalei	Kauai, Hawaii	Tropical wet	Coastal floodplain	0.82–3.09	100 years	Calhoun and Fletcher (1999)
Wainimala	Viti Levu, Fiji	Tropical wet	Floodplain	3.2	35 years	Terry *et al.* (2002b)
Falefa	Upolu, Samoa	Tropical wet	Floodplain	4.1	40 years	Terry *et al.* (2005)
Jourdain	Santo, Vanuatu	Tropical wet	Braidplain	4.8	40 years	Terry *et al.* (2006b)
Labasa	Vanua Levu, Fiji	Tropical wet	Floodplain	1.0–5.8	40 years	Terry *et al.* (2006c)
Short-term observations[a]						
Navua	Viti Levu, Fiji	Tropical wet	Delta	31	'Hurricane'-induced flood 1869	Derrick (1951)
Dumbéa	Grande Terre, New Caledonia	Tropical wet	Floodplain	0–10	TC Brenda flood 1968	Baltzer (1972)
several rivers on the plains	Guadalcanal, Solomon Islands	Tropical wet	Floodplains	30–50	TC Namu flood 1986	Trustrum *et al.* (1989)
Juruá	Brazil	Tropical wet	Floodplain	0–60	Annual flood 1985–86	Campbell *et al.* (1992)
Amazon	Brazil	Tropical wet	Floodplain	27–42	Recent annual floods	Mertes (1994)

[a]Short-term sedimentation rates are normally much higher than long-term rates, because recently laid sediments have not yet been compacted.

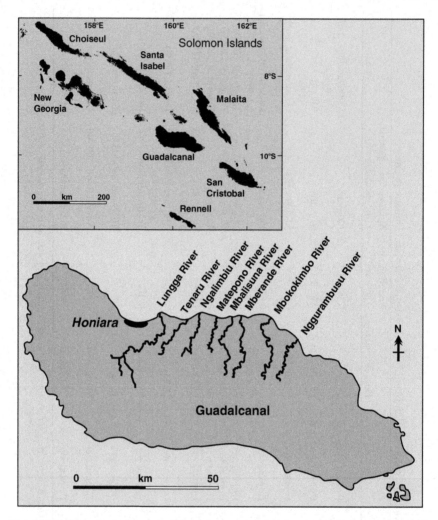

FIG. 10.12. Principal rivers draining the northern side of Guadalcanal island in Solomon Islands. Most rivers experienced devastating floods and unprecedented channel and valley aggradation on 18–19 May 1986, during Tropical Cyclone Namu. Adapted from Trustrum *et al.* (1989).

sedimentation that can be accomplished by a cyclone-generated flood. Most rivers and streams on Guadalcanal island experienced heavy aggradation. For example, 5 km above its estuary the Ngurambusu River aggraded by 4 m. In the river channels of upper watersheds, so-called *debris floods* were responsible for spectacular amounts of sedimentation. A debris flood is the term used to describe a process that is transitional between a debris flow

FIG. 10.13. Bridge over the Ngalimbiu River on Guadalcanal island in Solomon Islands, wrecked by floodwaters and sediments during Tropical Cyclone Namu on 18–19 May 1986. Photo courtesy of the Department of Mines and Energy, Solomon Islands.

TABLE 10.3. Extent of sediment deposition by river floods and landslides after Tropical Cyclone Namu in May 1986, in basins (from west to east) on Guadalcanal island in Solomon Islands

River basin	Basin area (km^2)	Percentage of basin affected by deposition (%)
Lungga	394	4
Tenaru	170	17
Ngalimbiu	262	18
Matepono	202	31
Mbalisuna	235	0
Mberande	234	5
Mbokokimbo	385	<1
Nggurambusa	205	0

River locations shown in Fig. 10.12. Summarised from Trustrum *et al.* (1989).

(a type of mass movement) and massive river bedload transport (a fluvial process), and exhibits features of both. In the highlands of the Mbalisuna basin, the pre-cyclone Namu channel of the Sutakiki River was infilled by coarse gravels to an astonishing thickness of up to 8 m (Fig. 10.14). Valembaimbai village was completely buried by gravels, killing 38 people,

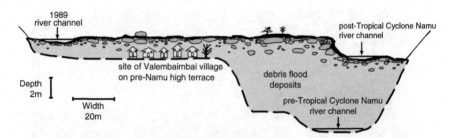

FIG. 10.14. Cross-section of the lower Sutakiki River, a main tributary of the Mbalisuna River on Guadalcanal island in Solomon Islands, before and after the flood generated by Tropical Cyclone Namu on 18–19 May 1986. Valembaimbai village was totally obliterated by bouldery debris-flood deposits. Adapted from Trustrum *et al.* (1989).

even though houses were situated upon a high terrace 7 m above the old Sutakiki channel. In the years since the TC Namu catastrophe, many rivers on Guadalcanal have been continually realigning their channels because of the geomorphic disequilibrium produced by these phenomenal debris floods.

References

Allan RJ, Nicholls N, Jones PD, Butterworth IJ (1991) A further extension of the Tahiti-Darwin SOI, early ENSO events and Darwin pressure. Journal of Climate 4:743–749

Allison MA, Kuehl SA, Martin TC, Hassan A (1998) The importance of floodplain sedimentation for river sediment budgets and terrigenous input to the oceans: insights from the Brahmaputra–Jamuna River. Geology 22:175–178

Anthes RA (1982) Tropical cyclones: their evolution, structure and effects. American Meteorological Society, Meteorological Monographs 19(41):208pp

Australian Bureau of Meteorology (2004) The South Pacific sea level and climate monitoring project. Monthly data report no 103, January 2004, 32pp

Australian Bureau of Meteorology (2005a) Accessed December 2005 from <http://www.bom.gov.au/climate/averages/climatology/tropical_cyclones/tropical_cycl.shtml>

Australian Bureau of Meteorology (2005b) Accessed October 2005 from <http://www.bom.gov.au/info/cyclone/>

Australian Bureau of Meteorology (2006) Accessed September 2006 from <http://www.bom.gov.au/weather/wa/cyclone/about/faq/faq_char_3.shtml>

Baines GBK, McLean RF (1976) Sequential studies of hurricane deposits evolution at Funafuti Atoll. Marine Geology 21:M1–M8

Baisyet PM (1993) Cyclone damage assessment of the Vaisigano watershed area. Department of Agriculture, Forests and Fisheries, Government of Samoa, United Nations Development Programme, and Food and Agriculture Organization, 15pp

Baltzer F (1972) Quelques effets sédimentologiques du cyclone Brenda dans la plaine alluviale de la Dumbéa (Côte ouest de la Nouvelle-Calédonie). Revue de Géomorphologie Dynamique 21:97–114

Barry RG, Chorley, RJ (1982) Atmosphere, weather and climate. 4th edn, Methuen, London, 407pp

Barry RG, Chorley RJ (1998) Atmosphere, weather and climate. 7th edn, Routledge, London, 409pp

Basher RE, Zheng X (1995) Tropical cyclones in the southwest Pacific: spatial patterns and relationships to Southern Oscillation and sea surface temperature. Journal of Climate 8:1249–1260

Bayliss-Smith TP (1988) The role of hurricanes in the development of reef islands, Ontong Java Atoll, Solomon Islands. The Geographical Journal 154:377–391

Bengtsson L, Botzet M, Esch M (1996) Will greenhouse gas-induced warming over the next 50 years lead to higher frequency and greater intensity of hurricanes? Tellus 48A:57–73c

Bettencourt S, Campbell J, de Wet N, Falkland A, Feresi J, Jones R, Kench P, Kenny G, King W, Lehodey P, Limalevu L, Raucher R, Taeuea T, Terry J, Teutabo N, Warrick R (2002) The impacts of climate change in Pacific island economies: policy and development implications. Asia Pacific Journal on Environment and Development 9:142–165

Blong R (1994) Natural perils and integrated hazard assessment in Fiji. Unpublished final report for Queensland Insurance (Fiji) Limited, National Insurance Company of New Zealand, and Fiji Reinsurance Corporation Limited

Bonell M, Gilmour DA (1978) The development of overland flow in a tropical rain-forest catchment. Journal of Hydrology 39:365–382

Bourrouilh-Le Jan FG, Talandier J (1985) Sédimentation et fracturation de haute energie en milieu récifal: tsunamis, ouragens et cyclones et leurs effets sur la sédimentologie et la géomorphologie d'un atoll: motu et hoa, à Rangiroa, Tuamotu. Marine Geology 67:263–333

Calhoun RS, Fletcher CH (1999) Measured and predicted sediment yield from a subtropical heavy rainfall, steep-sided river basin: Hanalei, Kauai, Hawaiian Islands. Geomorphology 30:213–226

Campbell DG, Stone JL, Rosas JL (1992) A comparison of the phytosociology and dynamics of three floodplain (Várzea) forests of known ages, Rio Juruá, western Brazilian Amazon. Botanical Journal of the Linnean Society 108:213–237

Chevalier J-P (1972) Observations sur les chenaux incomplets appelés hoa dans les atolls des Tuamotu. Proceedings of the Symposium on Corals and Coral Reefs 1969 (Mandapam Camp), pp. 477–488

Chu P-S (2004) ENSO and tropical cyclone activity. In: Murnane RJ, Liu K-B (eds) Hurricanes and typhoons: past, present and future. Columbia University Press, New York, pp. 297–332

Cloud PE (1952) Preliminary report on the geology and marine environments of Onotoa Atoll, Gilbert Islands. Atoll Research Bulletin 12:1–73

Congbin Fu, Diaz HF, Fletcher JO (1986) Characteristics of the response of the sea surface temperature in the central Pacific associated with warm episodes of the Southern Oscillation. Monthly Weather Review 114:1716–1738

Connell JH (1978) Diversity in tropical rain forests and coral reefs. Science 199:1302–1310

Cooper MJ (1966) Destruction of marine flora and fauna in Fiji caused by the hurricane of February 1965. Pacific Science 9:137–141

Crozier MJ, Howorth R, Grant IJ (1981) Landslide activity during cyclone Wally, Fiji: a case study of Wainitubatolu catchment. Pacific Viewpoint 22:69–88

D'Aubert AM (1994) Tropical cyclones and droughts in the Pacific Islands. Unpublished report. Department of Geography, The University of the South Pacific, Suva, Fiji

Derrick RA (1946) A history of Fiji, vol 1. Government Printing Department, Suva, Fiji

Derrick RA (1951) The Fiji Islands, a geographical handbook. Government Printing Department, Suva, Fiji, 334pp

Done TJ (1992) Effects of tropical cyclone waves on ecological and geomorphological structures on the Great Barrier Reef. Continental Shelf Research 12:223–233

Dvorak V (1984) Tropical cyclone intensity analysis using satellite data. NOAA technical report NESDIS 11. National Oceanic and Atmospheric Administration, Washington DC, 47pp

Editor of The Independent, March (1915) The Samoan incident. The Independent, newspaper editorial, New York

Elmqvist T, Rainey WE, Pierson ED, Cox PA (1994) Effects of tropical cyclones Ofa and Val on the structure of Samoan lowland rain forest. Biotropica 26:384–391

Emanuel KA (1987) The dependence of hurricane intensity on climate. Nature 326:483–485

Emanuel K (2004) Response of tropical cyclone activity to climate change: theoretical basis. In: Murnane RJ, Liu K-B (eds) Hurricanes and typhoons: past, present and future. Columbia University Press, New York, pp. 395–407

Emanuel K (2005) Increasing destructiveness of tropical cyclones over the past 30 years. Nature 436:686–688

Evans JL (1993) Sensitivity of tropical cyclone intensity to sea surface temperature. Journal of Climate 6:1133–1140

Feresi J, Kenny G, de Wet N, Limalevu L, Bhusan J, Ratukalou I (eds) (1999) Climate change vulnerability and adaptation assessment report for the Fiji Islands. Report for the South Pacific Regional Environment Programme (SPREP), Pacific Islands climate change assistance programme (PICCAP), Fiji Department of Environment, and the International Global Change Institute (IGCI). University of Waikato, New Zealand, 135pp

Fiji Meteorological Service (undated) List of floods occurring in the Fiji Islands between 1840 and 2000. Information sheet no 125. Climate Services Division, Nadi Airport, Fiji

Fiji Meteorological Service (1980) Cyclone Wally. Climate Services Division, Nadi Airport, Fiji, 10pp

Fiji Meteorological Service (1983) Annual weather summary 1982. Climate Services Division, Nadi Airport, Fiji

Fiji Meteorological Service (1986) Annual weather summary 1985. Climate Services Division, Nadi Airport, Fiji

Fiji Meteorological Service (1987) Annual weather summary 1986. Climate Services Division, Nadi Airport, Fiji

Fiji Meteorological Service (1990) Tropical Cyclone Ofa 31 January to 7 February 1990. Tropical cyclone report 90/4. Climate Services Division, Nadi Airport, Fiji, 13pp

Fiji Meteorological Service (1995) Annual weather summary 1994. Climate Services Division, Nadi Airport, Fiji

Fiji Meteorological Service (1997a) Preliminary report on Tropical Cyclone June, 3–5 May 1997. Climate Services Division, Nadi Airport, Fiji, 5pp

Fiji Meteorological Service (1997b) Preliminary report on Tropical Cyclone Gavin, 4–11 March 1997. Climate Services Division, Nadi Airport, Fiji, 4pp

Fiji Meteorological Service (1997c) Monthly weather summary for Fiji – January 1997. Climate Services Division, Nadi Airport, Fiji, 4pp

Fiji Meteorological Service (1997d) Tropical cyclone damages in Fiji (from 1939/40–96/97). Information sheet no 123. Climate Services Division, Nadi Airport, Fiji

Fiji Meteorological Service (2003) List of tropical cyclones in the south west Pacific 1969/70 – present. Information sheet no 121. Climate Services Division, Nadi Airport, Fiji

180   References

Fiji Meteorological Service (2004) Tropical Cyclone Heta (03F), 1–8 January 2004. Preliminary report. Climate Services Division, Nadi Airport, Fiji

Fiji Meteorological Service (2005) Personal communication with the FMS Director, Mr Rajendra Prasad.

Fiji Meteorological Service (2006) Accessed March 2006 from <http://www.met.gov.fj/services.html>

Fiji National Archives (1931) Hurricanes and foods, 1931. Fiji Legislative Council Paper no 23. Suva, Fiji

Fiji National Archives (1956) Tropical storm 29th–31st January 1956. Secretariat files F81/20/24. Suva, Fiji

Free M, Bister M, Emanuel K (2004) Potential intensity of tropical cyclones: comparison of results from radiosonde and reanalysis data. Journal of Climate 17:1722–1728

Garbell MA (1947) Tropical and equatorial meteorology. Pitman, New York, 237pp

Goodbred SL, Kuehl SA (1998) Floodplain processes in the Bengal basin and the storage of Ganges–Brahmaputra river sediment: an accretion study using ^{137}Cs and ^{210}Pb geochronology. Sedimentary Geology 121:239–258

Gray WM (1979) Hurricanes: their formation, structure and likely role in the tropical circulation. In: Shaw DB (ed) Meteorology over tropical oceans. Royal Meteorological Society, Bracknell, pp. 155–218

Greenbaum D, Bowker MR, Dau I, Dropsy H, Greally KB, McDonald AJW, Marsh SH, Northmore KJ, O'Connor EA, Prasad S, Tragheim DG (1995) Rapid methods of landslide hazard mapping: Fiji case study. British Geological Survey, technical report WC/95/28, 103pp

Guilcher A (1988) Coral reef geomorphology. Wiley, Chichester, 228pp

Hackman BD (1980) The geology of Guadalcanal, Solomon Islands. Institute of Geological Sciences, NERC overseas memoir, vol 6, 115pp

Harmelin-Vivien ML, Laboute P (1986) Catastrophic impact of hurricanes on atoll outer reef slopes in the Tuamotu (French Polynesia). Coral Reefs 5:55–62

Hasegawa A, Emori S (2005) Tropical cyclones and associated precipitation over the western North Pacific: T106 atmospheric GCM simulation for present-day and doubled CO2 climates. SOLA 1:145–148

Hastings PA (1990) Southern Oscillation influences on tropical cyclone activity in the Australian/south-west Pacific region. International Journal of Climatology 10:291–298

Hay J, Salinger J, Fitzharris B, Basher R (1993) Climatological seesaws in the southwest Pacific. Weather and Climate 13:9–21

Henderson-Sellers A, Zhang H, Berz G, Emanuel K, Gray W, Landsea C, Holland G, Lighthill J, Shieh S-L, Webster P, McGuffie K (1998) Tropical cyclones and global climate change: a post-IPCC assessment. Bulletin of the American Meteorological Society 79:19–38

Herwitz SR (1985) Interception storage capacities of tropical rainforest canopy trees. Journal of Hydrology 77:237–252

Herwitz SR (1986) Infiltration-excess caused by stemflow in a cyclone-prone tropical rainforest. Earth Surface Processes and Landforms 11:401–412

Hilton AC (1998) The influence of El Niño-Southern Oscillation (ENSO) on frequency and distribution of weather-related disasters in the Pacific Islands region. In: Terry JP (ed) Climate and environmental change in the Pacific. School of Social and Economic Development. The University of the South Pacific, Suva, Fiji, pp. 57–71

Holland G (1997) The maximum potential intensity of tropical cyclones. Journal of Atmospheric Science 54:2519–2541

Houghton JT, Ding Y, Griggs DJ, Nouger M, van der Linder PJ, Dai X, Maskell K, Johnson CA (eds) (2001) Climate change 2001: the scientific basis. Contribution of working group I to the third assessment report of the Intergovernmental Panel on Climate Change (IPCC). Cambridge University Press, Cambridge, 478pp

Howorth R, Prasad S (1981) Morphology of typical landslides which occurred Good Friday 1980 in the Serua Hills, Viti Levu, Fiji. Fiji Mineral Resources Department, report no 28, Suva, Fiji, 29pp

Howorth R, Crozier MJ, Grant IJ (1981) Effects of Tropical Cyclone Wally in southeast Viti Levu, Fiji, Easter 1980. Search 12:41–43

Hulme M, Viner D (1998) A climate change scenario for the tropics. Climatic Change 39:145–176

Intes A, Caillart B (1994) Environment and biota of the Tikehau Atoll (Tuamotu Archipelago French Polynesia), Part I. Atoll Research Bulletin 415

IPCC (2001) Intergovernmental panel on climate change. Climate change 2001: the scientific basis. Contribution of working group I to the third assessment report of the IPCC. Houghton JT, Ding Y, Griggs DJ, Noguer M, van der Linden PJ, Dai X, Maskell K, Johnson CA (eds). Cambridge University Press, Cambridge and New York, 881pp

Jackson IJ (1971) Problems of throughfall and interception assessment under tropical forest. Journal of Hydrology 12:234–254

Jelesnianski CP (1972) SPLASH (special program to list amplitudes of surges from hurricanes) I. Landfall storms. NOAA technical memorandum NWS TDL-46. National Weather Service Systems Development Office, Silver Spring, Maryland, 56pp

JICA (1997) Japan International Co-operation Agency/Ministry of Agriculture, Fisheries and Forests Fiji. Interim Report, October 1997. The study on watershed management and flood control for the four major Viti Levu rivers in the Republic of Fiji

JTWC (2006) Joint Typhoon Warning Centre. US Naval Pacific Meteorology and Oceanography Center. Accessed September 2006 from <http://www.npmoc.navy.mil/jtwc/menu/JTFAQ.html>

Karoly DJ, Wu Q (2005) Detection of regional surface temperature trends. Journal of Climate 18:4337–4343

Kerr IS (1976) Tropical storms and hurricanes in the southwest Pacific, November 1939 to April 1969. New Zealand Meteorological Service, miscellaneous publication no 148, Wellington, 114pp

Klotzbach PJ (2006) Trends in global tropical cyclone activity over the past twenty years (1986–2005). Geophysical Research Letters 33:L10805, doi:10.1029/2006GL025881

Knutson TR, Tuleya RE (2004) Impact of CO_2-induced warming on simulated hurricane intensity and precipitation: sensitivity to the choice of climate model and convective parameterization. Journal of Climate 17:3477–3494

Knutson TR, Tuleya RE, Shen W, Ginis I (2001) Impact of CO_2-induced warming as simulated using the GFDL hurricane prediction system. Climate Dynamics 15:03–519

Knutson TR, Tuleya RE, Shen W, Ginis I (2004) Impact of climate change on hurricane intensities as simulated using a regional nested high-resolution models. In: Murnane RJ, Liu K-B (eds) Hurricanes and Typhoons: Past, Present and Future. Columbia University Press, New York, pp. 408–439

Kostaschuk R, Terry J, Raj R (2001) The impact of tropical cyclones on river floods in Fiji. Hydrological Sciences Journal 46:435–450

Kostaschuk R, Terry J, Raj R (2003) Suspended sediment transport during tropical cyclone floods in Fiji. Hydrological Processes 17:1149–1164

Kostaschuk RA, Terry JP, Wotling G (2006) Tropical storms and associated flood risk on Grande Terre, New Caledonia. In: Sethaputra S, Promma K (eds) Proceedings of the international symposium on managing water supply for growing demand, 16–20 October 2006, Bangkok, Thailand. UNESCO International Hydrological Programme, technical documents in hydrology no 6, pp. 207–210

Krishna R (1984) Tropical cyclones. Fiji Meteorological Service, publication no 4

Landsea CW, Harper BA, Hoarau K, Knaff JA (2006) Can we detect trends in extreme tropical cyclones? Science 313:452–454

Lang M, Ouarda TBMJ, Bobée B (1999) Towards operational guidelines for over-threshold modeling. Journal of Hydrology 225:103–117

Lawson T (1991) South-east Viti Levu landslide project. Submission of staffing and financial assistance. Unpublished internal report. Fiji Mineral Resources Department, Suva

Lawson T (1993) Summary of the south-east Viti Levu landslide project – preliminary study. Fiji Mineral Resources Department, Suva

Liu K-B (2004) Paleotempestology: principles, methods, and examples from Gulf Coast lake sediments. In: Murnane RJ, Liu K-B (eds) Hurricanes and typhoons: past, present and future. Columbia University Press, New York, pp. 13–57

Liu G, Curry JA, Clayson CA (1995) Study of tropical cyclogenesis using satellite data. Meteorology and Atmospheric Physics 56:111–123

Lovell E (2006) School of Marine Studies, The University of the South Pacific, Suva, Fiji, personal communication

Madsen H, Rasmussen PF, Rosbjerg D (1997) Comparison of annual maximum series and partial duration series methods for modeling extreme hydrologic events. Water Resources Research 33:747–757

Maragos JE, Baines GBK, Beveridge PJ (1973) Tropical Cyclone Bebe creates a new land formation on Funafuti Atoll. Science 181:1161–1164

McGregor GR, Nieuwolt S (1998) Tropical climatology: an introduction to the climates of the low latitudes. 2nd edn, Wiley, Chichester, UK, 352pp

McKee ED (1959) Storm sediments on a Pacific atoll. Journal of Sedimentary Petrology 29:354–364

McLean RF, Hosking PL (1991) Geomorphology of reef islands and atoll motu in Tuvalu. South Pacific Journal of Natural Science 11:167–189

Mertes LAK (1994) Rates of flood-plain sedimentation on the central Amazon River. Geology 22:171–174

Morrison RJ (1981) Factors determining the extent of soil erosion in Fiji. Institute of Natural Resources, The University of the South Pacific, environmental studies report no 7, 18pp

Murnane RJ (2004) The importance of best-track data for understanding the past, present and future of hurricanes and typhoons. In: Murnane RJ, Liu K-B (eds) Hurricanes and typhoons: past, present and future. Columbia University Press, New York, pp. 249–266

Murray WE, Terry JP (2004) Niue's place in the Pacific. In: Terry JP, Murray WE (eds) Niue Island, geographical perspectives on the Rock of Polynesia. International Scientific Council for Island Development, UNESCO, Paris, pp. 9–29

NASA (2006) National Aeronautics and Space Administration. Earth Observatory. Accessed March 2006 from <http://earthobservatory.nasa.gov/Library/Hurricanes/Images/storm_surge.gif>

National Disaster Council (1986) Report on the post-cyclone survey. Solomon Islands Government, Honiara, 27pp

Newell ND, Bloom AL (1970) The reef flat and 'two-meter eustatic terrace' of some Pacific atolls. Geological Society of America Bulletin 81:1881–1894

NIWA (2003) National Institute of Water and Atmospheric Research, New Zealand. The island climate update, February 2003, no 29, 4pp

NOAA (2006a) National Oceanic and Atmospheric Administration. Accessed May 2006 from <http://www.cdc.noaa.gov/ENSO/enso.faq.html#11>

NOAA (2006b) National Oceanic and Atmospheric Administration. Accessed January 2006 from <http://www.aoml.noaa.gov/hrd/tcfaq/D6.html>

NOAA (2006c) National Oceanic and Atmospheric Administration. Accessed May 2006 from <http://www.aoml.noaa.gov/hrd/tcfaq/D1.html>

Nott JF (2003) Intensity of prehistoric tropical cyclones. Journal of Geophysical Research 108(D7):4212, p. ACL 5-1–5-7

Nott JF (2004) Paleotempestology: the study of prehistoric tropical cyclones – a review and implications for hazard assessment. Environment International 30:433–447

Nott JF (2006) Tropical cyclones and the evolution of the sedimentary coast of northern Australia. Journal of Coastal Research 22:49–62

NSR Environmental Consultants (1994) Namosi prospect environmental program. Progress data report, August 1992–December 1993, Hawthorn, Australia

Nunn PD (1994) Oceanic islands. Blackwell, Oxford, 413pp

OCHA (2005) United Nations Office for the Coordination of Humanitarian Affairs. OCHA situation report no 5, Tokelau – Tropical Cyclone Percy, 1 April 2005. Accessed September 2006 from <http://www.reliefweb.int/rw/RWB.NSF/ db900SID/ SNAO-6B2KVA?OpenDocument>

Ollier CD, Terry JP (1999) The volcanic geomorphology of northern Viti Levu. Australian Journal of Earth Science 46:515–522

Oouchi K, Yoshimura J, Yoshimura H, Mizuta R, Kusunoki S, Noda A (2006) Tropical cyclone climatology in a global-warming climate as simulated in a 20 km-mesh global atmospheric model: frequency, and wind intensity analysis. Journal of the Meteorological Society of Japan 84:259–276

ORE (2002) Observatoire de la Ressource en Eau. Water management in New Caledonia. Government of New Caledonia, Nouméa, 16pp

Parry JT (1987) The Sigatoka valley, pathway to prehistory. Bulletin of the Fiji Museum no 9, Suva, 157pp

Pielke RA, Pielke RA (1997) Hurricanes: their nature and impacts on society. Wiley, Chichester, 279pp

Pirazzoli PA, Delibrias G, Montaggioni LF, Saliège JF, Vergnaud-Grazzini C (1987) Vitesse de croissance latérale des platiers et évolution morphologique récente de l'atoll de Reao, îles Tuamotu, Polynésie française. Annales de l'Institut Océanographipe 63:57–68

Public Works Department (2000) Major floods in western towns. Fiji PWD Hydrology Division, unpublished miscellaneous flood records, Suva

Raper S (1992) Observational data on the relationships between climatic change and the frequency and magnitude of severe tropical storms. In: Warwick RA, Barrow EM,

Wigley TML (eds) Climate and sea level change: observations, projections and implications. Cambridge University Press, Cambridge, pp. 192–212

Raucher RS (2000) Economic implications of climate change in two Pacific Island country locations. Case illustration of Tarawa, Kiribati and Viti Levu. Prepared for the World Bank by Stratus Consulting, Boulder

Rearic DM (1990) Survey of Cyclone Ofa damage to the northern coast of Upolu, Western Samoa. South Pacific Applied Geoscience Commission, Suva. SOPAC technical report no 104, 37pp

Revell CG (1981) Tropical Cyclones in the southwest Pacific, November 1969 to April 1979. New Zealand Meteorological Service, miscellaneous publication no 170, Wellington, 53pp

Revell CG, Goulter SW (1986a) South Pacific tropical cyclones and the Southern Oscillation. Monthly Weather Review 114:1138–1145

Revell CG, Goulter SW (1986b) Lagged relations between the Southern Oscillation and numbers of tropical cyclones in the South Pacific region. Monthly Weather Review 114:2669–2670

Riehl H (1954) Tropical meteorology. McGraw-Hill, New York, 392pp

Rodda P (1990) Rate of movement of meanders along the lower Wainimala, and heights of alluvial terraces. Fiji Mineral Resources Department, note BP1/85, 15pp

Ropelewski CF, Jones PD (1987) An extension of the Tahiti-Darwin Southern Oscillation Index. Monthly Weather Review 115:2161–2165

Rouse WC, Reading A (1985) Soil mechanics and natural slope stability. In: Richards KS, Arnett RR, Ellis S (eds) Geomorphology and soils. Allen and Unwin, London, pp. 159–178

Royer J-F, Chauvin F, Timbal B, Araspin P, Grimal D (1998) A GCM study of the impact of greenhouse gas increase on the frequency of occurrence of tropical cyclones. Climatic Change 38:307–343

Salinger MJ, Basher RE, Fitzharris BB, Hay JE, Jones PD, MacVeigh JP, Schmidely-Leleu L (1995) Climate trends in the south-west Pacific. International Journal of Climatology 15:285–302

Salvat F, Salvat B (1992) Nukutipipi Atoll, Tuamotu Archipelago; geomorphology, land and marine flora and fauna and interrelationships. Atoll Research Bulletin 357

Salvat B, Chevalier J-P, Richard G, Poli G, Bagnis R (1977) Geomorphology and biology of Taiaro Atoll, Tuamotu Archipelago. Proceedings of the 3rd international coral reef symposium, vol 2, pp. 289–295

Santer BD, Wigley TML, Gleckler PJ, Bonfils C, Wehner MF, Achuta Rao K, Barnett TP, Boyle JS, Brüggemann W, Fiorino M, Gillett N, Hansen JE, Jones PD, Klein SA, Meehl GA, Raper SCB, Reynolds RW, Taylor KE, Washington WM (2006) Forced and unforced ocean temperature changes in Atlantic and Pacific tropical cyclogenesis regions. Proceedings of the National Academy of Science 103:13905–13910

Schofield JC (1959) The geology and hydrology of Niue Island, South Pacific. New Zealand Geological Survey, bulletin no. 63, 27pp

Scoffin TP (1993) The geological effects of hurricanes on coral reefs and the interpretation of storm deposits. Coral Reefs 12:203–221

Simpson RH, Riehl H (1981) The hurricane and its impact. Louisiana State University Press, Baton Rouge, 398pp

Stephens PR, Trustrum NA, Fletcher JR (1986) Reconnaissance of the physical impact of Cyclone Namu – Guadalcanal and Malaita, Solomon Islands. Soil

Conservation Centre, Ministry of Works and Development, Palmerston North, New Zealand, unpublished internal report no 178

Stevenson RL (1915) A footnote to history. March 1915. The Independent, New York

Stoddart DR (1971) Coral reefs and islands and catastrophic storms. In: Steers JA (ed) Applied coastal geomorphology. MacMillan, London, pp. 155–197

Stoddart DR (1975) Almost-atoll of Aitutaki: geomorphology of reefs and islands. Atoll Research Bulletin 190:31–57

Stoddart DR (1985) Hurricane effects on coral reefs: conclusion. Proceedings of the fifth international coral reef symposium, vol 3, pp. 349–350

Stoddart DR, Fosberg RF (1994) The hoa of Hull Atoll and the problem of hoa. Atoll Research Bulletin 394, 26pp

Stoddart DR, Walsh RPD (1992) Environmental variability and environmental extremes as factors in the island ecosystem. Atoll Research Bulletin 356, 71pp

Strahler A, Strahler A (2006) Introducing physical geography. 4th edn, Wiley, 728pp

Terry JP (2002) Water resources, climate variability and climate change in Fiji. Asia Pacific Journal on Environment and Development 9:86–120

Terry JP (2004a) Climatic hazards facing Niue. In: Terry JP, Murray WE (eds) Niue Island, geographical perspectives on the Rock of Polynesia. International Scientific Council for Island Development, UNESCO, Paris, pp. 113–124

Terry JP (2004b) Shoreline erosion on a low coral island in Fiji – causes and consequences. South Pacific Studies 24:55–66

Terry JP (2005) Hazard warning! Hydrological responses in the Fiji Islands to climate variability and severe meteorological events. In: Franks S, Wagener T, Bøgh E, Gupta HV, Bastidas L, Nobre C, de Oliveira Galvão C (eds) Regional hydrological impacts of climatic change – hydroclimatic variability. International Association of Hydrological Sciences, publication no 296, pp. 33–41

Terry JP, Kostaschuk RA (2001) Rapid rates of channel migration in a Pacific island river. Journal of Pacific Studies 25:277–289

Terry JP, Raj R (1999) Island environment and landscape responses to 1997 tropical cyclones in Fiji. Pacific Science 13:257–272

Terry JP, Raj R (2002) The 1997–98 El Niño and drought in the Fiji Islands. In: Hydrology and water management in the humid tropics. UNESCO International Hydrology Programme-V, technical documents in hydrology no 52, Paris, pp. 80–93

Terry JP, Ollier CD, Pain CF (2002a) Geomorphological evolution of the Navua River, Fiji. Physical Geography 23:418–426

Terry JP, Garimella S, Kostaschuk RA (2002b) Rates of floodplain accretion in a tropical island river system impacted by cyclones and large floods. Geomorphology 42:171–183

Terry JP, McGree S, Raj R (2004) The exceptional floods on Vanua Levu island, Fiji, during Tropical Cyclone Ami in January 2003. Journal of Natural Disaster Science 26:27–36

Terry JP, Kostaschuk RA, Garimella S (2005) Sediment deposition rate in the Falefa River basin, Upolu island, Samoa. Journal of Environmental Radioactivity 86:45–63

Terry JP, Kisun P, Qareqare, Rajan, J (2006a) Lagoon degradation and management in Yanuca channel on the Coral Coast, Fiji. South Pacific Journal of Natural Science 24:1–13

Terry JP, Garimella S, Kostaschuk RA, Ratiram A (2006b) Investigating the evolution of the Jourdain River braidplain on Santo island, Vanuatu. Australia and New Zealand Geomorphology Group, occasional paper no 4, p. 32

Terry JP, Lal R, Garimella S (2006c) Factors influencing active sedimentation in the Labasa River valley on Vanua Levu island in Fiji, measured from caesium-137 and lead-210 profiles. In: Collen J, Bukarau L (eds) South Pacific Applied Geoscience Commission, Suva. SOPAC miscellaneous report no 621, pp. 52–53

Timmermann A, Oberhuber J, Bacher A, Esch M, Latif M, Roeckner E (1999) Increased El Niño frequency in a climate model forced by future greenhouse warming. Nature 398:694–697

Tonkin H, Holland GJ, Holbrook N, Henderson-Sellers A (2000) An evaluation of thermodynamic estimates of climatological maximum potential tropical cyclone intensity. Monthly Weather Review 128:746–762

Trenberth K, Hoar TJ (1997) El Niño and climate change. Geophysical Research Letters 24:3057–3060

Trustrum NA, Whitehouse IE, Blaschke PM (1989) Flood and landslide hazard, northern Guadalcanal, Solomon Islands. Department of Scientific and Industrial Research, New Zealand. Unpublished report for United Nations Technical Cooperation for Development. New York, 6/89 SOI/87/001.43

Trustrum NA, Whitehouse IE, Blaschke PM, Stephens PR (1990) Flood and landslide hazard mapping, Solomon Islands. In: Ziemer RR, O'Loughlin CL, Hamilton LS (eds) Research needs and applications to reduce erosion and sedimentation in tropical steeplands. International Association of Hydrological Sciences, publication no 197, pp. 138–146

Vaithiyanathan P, Ramanathan AL, Subramanian V (1988) Erosion, transport and deposition of sediments by the tropical rivers of India. In: Bordas MP, Walling DE (eds) Sediment budgets. International Association of Hydrological Sciences, publication no 174, pp. 561–574

Velden C, Harper B, Wells F, Beven II JL, Zehr R, Olander T, Mayfield M, Guard C, Lander M, Edson R, Avila L, Burton A, Turk M, Kikuchi A, Christian A, Caroff P, McCrone P (2006) The Dvorak tropical cyclone intensity estimation technique: a satellite-based method that has endured for over 30 years. Bulletin of the American Meteorological Society 87:1195–1210

Vincent DG (1994) The South Pacific Convergence Zone (SPCZ): a review. Monthly Weather Review 112:1949–1970

Visher SS (1925) Tropical cyclones of the Pacific. Bulletin no 20 of the Bishop Museum, Honolulu, 163pp

Walsh RPD (1977) Changes in the tracks and frequency of tropical cyclones in the Lesser Antilles from 1650 to 1975 and some geomorphological and ecological implications. Swansea Geographer15:4–11

Walsh RPD (1980) Runoff processes and models in the humid tropics. Zeitschrift für Geomorphologie, NF supplement band 36:176–202

Walsh RPD (1982) A provisional study of the effects of hurricanes David and Fredrick in 1979 on the terrestrial environment of Dominica, West Indies. Swansea Geographer 19:28–35

Walsh KJE, Katzfey JJ (2000) The impact of climate change on the poleward movement of tropical cyclone-like vortices in a regional climate model. Journal of Climate 13:1116–1132

Walsh K, Pittock AB (1998) Potential changes in tropical storms, hurricanes, and extreme rainfall events as a result of climate change. Climatic Change 39:199–213

Walsh KJE, Ryan BF (2000) Tropical cyclone intensity increase near Australia as a result of climate change. Journal of Climate 13:3237–3254

Webster PJ, Holland GL, Curry JA, Chang H-R (2005) Changes in tropical cyclone number, duration, and intensity in a warming environment. Science 309:1844–1846

Wells S, Hanna, N (1992) The Greenpeace book of coral reefs. Blandford, London, 160pp

Williams J (1837) A narrative of missionary enterprises in the south sea islands. Appleton D and Co, 525pp

WMO (2004) World Meteorological Organization. Tropical cyclone operational plan for the South Pacific and south-east Indian Ocean. WMO technical document no 292, Geneva

WMO (2005) World Meteorological Organization. Fact sheet tropical cyclone names. Accessed December 2006 from <http://www.wmo.ch/web/www/TCP/TCnames 2004–2009.pdf>, 7pp

WMO (2006a) World Meteorological Organization. Accessed January 2006 from <http://www.wmo.ch/web/www/TCP/rsmcs.html>

WMO (2006b) World Meteorological Organization. Statement on tropical cyclones and climate Change. WMO international workshop on tropical cyclones, IWTC-6, San Jose, Costa Rica, November 2006, 13pp

WMO (2006c) World Meteorological Organization. Summary statement on tropical cyclones and climate change. WMO international workshop on tropical cyclones, IWTC-6, San Jose, Costa Rica, November 2006, 1pp

Wood-Jones F (1910) Coral and atolls. A history and description of the Keeling-Cocos Islands, with an account of their fauna and flora, and a discussion of the method of development and transformation of coral structures in general. Reeve, London, 392pp

Woodroffe CD (1983) The impact of Cyclone Isaac on the coast of Tonga. Pacific Science 37:181–210

World Bank (2000) Cities, seas and storms: managing change in Pacific Island economies, vol IV. Adapting to climate change (November 30, 2000 draft). East Asia and Pacific Region, Papua New Guinea and Pacific Islands Country Management Unit, World Bank, Washington DC, 48pp

Yeo SW (1998) Natural and human controls on flooding in the Ba River valley, Fiji. Unpublished PhD thesis, School of Earth Sciences, Macquarie University, Sydney

Yeo SW (2000) Ba community flood preparedness project: final report. South Pacific Applied Geoscience Commission, Suva. SOPAC technical report no 309, 37pp

Zann L (1991) Effects of Cyclone Ofa on the fisheries and coral reefs of Upolu, Western Samoa in 1990. Unpublished report prepared for the Government of Western Samoa, Food and Agricultural Organization of the United Nations. SAM/89/002 Technical Report no 2

Appendix

APPENDIX 1. List of tropical cyclones in the South Pacific[a] from the 1969–1970 season until the end of the 2005–2006 season. Updated from original source (Fiji Meteorological Service 2003).

Season	Name of cyclone[b]	Duration	Maximum intensity	Places affected[c]
1969–1970	Ada	03–19 Jan	Hurricane	Australia, New Caledonia
	Dolly	11–25 Feb	Hurricane	New Caledonia, Vanuatu, Tonga, Niue, Samoa, Southern Cook Islands, French Polynesia
	Dawn	12–19 Feb	Storm	New Caledonia
	Emma	27 Feb–06 Mar	Hurricane	Cook Islands, Society Islands, Austral Islands
	Gillian	08–11 Apr	Storm	French Polynesia
	Helen	13–16 Apr	Gale	Tonga
	Isa	14–18 Apr	Gale	Tonga, Wallis and Futuna, Solomon Islands
1970–1971	Odile	06–08 Dec	Gale	Wallis and Futuna
	Priscilla	16–19 Dec	Gale	Fiji
	Rosie	30 Dec–04 Jan	Storm	Vanuatu, New Caledonia
	Dora	11–17 Feb	Gale	Australia, New Caledonia
	Gertie	12–15 Feb	Storm	Australia
	Ida	17–23 Feb	Storm	New Caledonia
	Fiona	20 Feb–01 Mar	Gale	Australia, New Caledonia
	Lena	14–23 Mar	Storm	Australia, New Caledonia
1971–1972	Unnamed	05–12 Nov	Storm	New Caledonia
	Ursula	03–15 Dec	Hurricane	Vanuatu, Norfolk Island
	Vivienne	17–19 Dec	Gale	Cook Islands, Austral Islands
	Althea	21–30 Dec	Hurricane	Australia, New Caledonia
	Carlotta	08–26 Jan	Hurricane	Solomon Islands, Vanuatu, New Caledonia
	Unnamed	18–24 Jan	Gale	Tonga, Niue
	Daisy	05–16 Feb	Hurricane	Australia
	Wendy	29 Feb–09 Mar	Hurricane	Tuvalu, Vanuatu, New Caledonia, Australia
	Yolande	17–23 Mar	Hurricane	Vanuatu, New Caledonia

(continued)

APPENDIX 1. (continued)

Season	Name of cyclone[b]	Duration	Maximum intensity	Places affected[c]
	Agatha	22–28 Mar	Hurricane	Southern Cook Islands, French Polynesia
	Emily	29 Mar–04 Apr	Hurricane	Australia, New Caledonia
	Gail	12–20 Apr	Hurricane	Australia, Vanuatu, New Caledonia
	Faith	14–22 Apr	Gale	Australia, Vanuatu, New Caledonia
	Hannah	08–12 May	Hurricane	Australia, Papua New Guinea
	Ida	30 May–05 Jun	Hurricane	Papua New Guinea, New Caledonia
1972–1973	Bebe	19–28 Oct	Hurricane	Fiji, Tuvalu, Tonga
	Collette	02–03 Nov	Gale	Wallis, Tonga
	Diana	08–18 Dec	Hurricane	Vanuatu, New Caledonia
	Felicity	14–18 Jan	Gale	Southern Cook Islands, French Polynesia
	Glenda	31 Jan–01 Feb	Gale	Southern Cook Islands, French Polynesia
	Henrietta	28 Feb–02 Mar	Gale	Fiji
	Elenore	31 Jan–07 Feb	Storm	Samoa, Tonga, Niue, Kermadec Islands
	Kirsty	25 Feb–01 Mar	Gale	No land areas
	Madge	02–03 Mar	Gale	No land areas
	Juliette	02–06 Apr	Storm	Fiji, Tonga
1973–1974	Unnamed	07–11 Nov	Gale	Niue
	Lottie	05–12 Dec	Hurricane	New Caledonia, Vanuatu, Fiji, Tonga
	Una	16–19 Dec	Gale	Australia
	Monica	17–20 Jan	Gale	Vanuatu, Loyalty Islands, Kermadec Islands
	Nessie	19–22 Jan	Gale	No land areas
	Vera	19–22 Jan	Storm	No land areas
	Wanda	23–25 Jan	Gale	Australia
	Pam	30 Jan–07 Feb	Hurricane	Futuna, Rotuma, Vanuatu, New Caledonia
	Zoe	05 Mar–02 Apr	Storm	Australia
	Alice	20–30 Mar	Storm	Lord Howe Island
	Tina	24–28 Apr	Storm	Fiji, Niue
1974–1975	Flora	13–21 Jan	Storm	Vanuatu
	Gloria	16–20 Jan	Storm	New Caledonia
	Val	29 Jan–05 Feb	Hurricane	Fiji, Tonga, Samoa, Wallis and Futuna
	Alison	04–12 Mar	Hurricane	Vanuatu, New Caledonia
	Betty	30 Apr–12 May	Hurricane	Vanuatu, Fiji
1975–1976	David	11–21 Jan	Hurricane	Vanuatu, New Caledonia, Australia
	Elsa	21–26 Jan	Storm	Vanuatu, New Caledonia
	Alan	31 Jan–01 Feb	Storm	New Caledonia, Australia
	Frances	03–10 Feb	Hurricane	French Polynesia

(continued)

APPENDIX 1. (continued)

Season	Name of cyclone[b]	Duration	Maximum intensity	Places affected[c]
	Beth	15–22 Feb	Hurricane	Australia
	Colin	26 Feb–07 Mar	Hurricane	Louisiade Archipelago
	Dawn	05–06 Mar	Gale	Australia
	Hope	11–19 Mar	Gale	Vanuatu, New Caledonia, Lord Howe Island
	Jan (Ian)	16–21 Apr	Storm	New Caledonia, Vanuatu, Kermadec Islands
	Watorea	25–30 Apr	Hurricane	Louisiade Archipelago, Australia, New Zealand
1976–1977	Kim	09–13 Dec	Storm	Samoa, Southern Cook Islands, French Polynesia
	Laurie	11–12 Dec	Storm	Samoa, Southern Cook Islands, French Polynesia
	Marion	12–21 Jan	Storm	New Caledonia, Vanuatu, Kermadec Islands
	June	18–25 Jan	Hurricane	Vanuatu
	Unnamed	03–09 Feb	Gale	Tonga
	Miles	11–12 Feb	Storm	New Caledonia
	Unnamed	20–24 Feb	Gale	Southern Cook Islands
	Norman	09–24 Mar	Storm	New Caledonia, Vanuatu
	Pat	15–18 Mar	Storm	Tonga, Niue
	Robert	16–22 Apr	Hurricane	Cook Islands, French Polynesia
1977–1978	Tom	06–17 Nov	Gale	Louisiade Archipelago
	Steve	24 Nov–04 Dec	Hurricane	Tuvalu
	Tessa	06–09 Dec	Gale	French Polynesia, Northern Cook Islands
	Anne	23–31 Dec	Storm	Futuna, Tonga, Fiji, Niue
	Bob	31 Jan–12 Feb	Hurricane	Vanuatu, Fiji
	Charles	14–28 Feb	Hurricane	Southern Cook Islands, French Polynesia
	Diana	15–22 Feb	Storm	Northern Cook Islands, Society Islands
	Ernie	16–23 Feb	Storm	Fiji, Tonga
	Gwen	01–03 Mar	Gale	No land areas
	Hal	07–22 Apr	Storm	New Zealand
1978–1979	Fay	27–31 Dec	Storm	Tuvalu, Fiji
	Gordon	03–12 Jan	Hurricane	Tuvalu, Vanuatu, New Caledonia
	Henry	29 Jan–05 Feb	Storm	New Caledonia, Vanuatu, New Zealand
	Kerry	14 Feb–07 Mar	Hurricane	Solomon Islands, Australia
	Leslie	21–23 Feb	Storm	Tonga
	Meli	24–31 Mar	Hurricane	Fiji, Tonga, Kermadec Islands
1979–1980	Ofa	10–15 Dec	Storm	Wallis, Tonga, Niue
	Peni	02–06 Jan	Hurricane	Wallis, Niue, Fiji
	Paul	07–12 Jan	Storm	No land areas
	Rae	02–05 Feb	Gale	New Caledonia, Vanuatu

(continued)

APPENDIX 1. (continued)

Season	Name of cyclone[b]	Duration	Maximum intensity	Places affected[c]
	Ruth	11–17 Feb	Gale	No land areas
	Leslie	21–23 Feb	Storm	French Polynesia
	Simon	22 Feb–03 Mar	Hurricane	No land areas
	Sina	11–16 Mar	Hurricane	Australia, New Zealand
	Tia	22–27 Mar	Storm	New Caledonia, Fiji, New Zealand
	Val	25–28 Mar	Storm	Futuna
	Wally	02–07 Apr	Gale	Fiji
1980–1981	Diola	27–30 Nov	Gale	French Polynesia
	Arthur	11–17 Jan	Hurricane	Fiji
	Betsy	30 Jan–03 Feb	Gale	Niue, Loyalty Islands, Tonga
	Eddie	08–10 Feb	Storm	Gulf of Carpentaria
	Cliff	08–15 Feb	Hurricane	New Caledonia, Vanuatu, Niue, Tonga
	Unnamed	16–20 Feb	Gale	Tonga, Niue
	Daman	20–24 Feb	Storm	Tonga, Southern Cook Islands
	Unnamed	23–28 Feb	Gale	Southern Cook Islands
	Unnamed	23 Feb–06 Mar	Gale	Australia
	Freda	26 Feb–09 Mar	Hurricane	Australia, New Caledonia
	Esau	01–05 Mar	Storm	French Polynesia
	Tahmar	09–13 Mar	Hurricane	French Polynesia
	Fran	17–24 Mar	Storm	Southern Cook Islands, French Polynesia, New Caledonia, Samoa
	Unnamed	27–31 Mar	Gale	No land areas
1981–1982	Gyan	19–29 Dec	Hurricane	Vanuatu, New Caledonia
	Abigail	23 Jan–07 Feb	Hurricane	New Caledonia
	Hettie	24 Jan–06 Feb	Hurricane	Fiji, Vanuatu, New Caledonia
	Isaac	27 Feb–05 Mar	Hurricane	Tonga
	Bernie	01–14 Apr	Hurricane	Solomon Islands, New Caledonia
	Dominic	09–13 Apr	Gale	No land areas
	Claudia	13–17 May	Gale	Solomon Islands, New Caledonia
1982–1983	Joti	31 Oct–07 Nov	Storm	Vanuatu, New Caledonia
	Kina	07–12 Nov	Gale	Vanuatu, New Caledonia
	Lisa	10–18 Dec	Storm	French Polynesia
	Des	15–22 Jan	Storm	New Caledonia
	Mark	20 Jan–01 Feb	Hurricane	New Caledonia, Vanuatu
	Nano	21–26 Jan	Hurricane	French Polynesia
	Elinor	12 Feb–03 Mar	Hurricane	New Caledonia
	Nisha (Orama)	20–27 Feb	Hurricane	French Polynesia
	Oscar	23 Feb–06 Mar	Hurricane	Fiji, Tahiti
	Prema	25 Feb–06 Mar	Gale	Northern Cook Islands, French Polynesia
	Rewa	07–16 Mar	Hurricane	French Polynesia
	Saba	20–24 Mar	Storm	French Polynesia
	Sarah	23 Mar–03 Apr	Hurricane	Fiji, Tahiti, Society Islands

(continued)

APPENDIX 1. (continued)

Season	Name of cyclone[b]	Duration	Maximum intensity	Places affected[c]
	Tomasi	28 Mar–05 Apr	Hurricane	Niue, French Polynesia
	Veena	08–14 Apr	Hurricane	Niue, French Polynesia
	William	15–23 Apr	Hurricane	Tahiti
1983–1984	Fritz	10–14 Dec	Gale	No land areas
	Atu	27–30 Dec	Gale	New Caledonia
	Grace	11–19 Jan	Hurricane	New Caledonia
	Beti	01–05 Feb	Hurricane	Vanuatu, New Caledonia
	Harvey	04–10 Feb	Storm	New Caledonia, Loyalty Islands
	Ingrid	20–27 Feb	Hurricane	New Caledonia
	Unnamed	20–24 Feb	Gale	New Caledonia
	Cyril	16–21 Mar	Gale	Fiji, Loyalty Islands
	Unnamed	23–30 Mar	Gale	Tonga
	Lance	03–08 Apr	Storm	Wallis, Tonga
1984–1985	Monica	25 Dec–03 Jan	Storm	New Caledonia
	Unnamed	26–28 Dec	Gale	Tuvalu
	Drena	09–16 Jan	Gale	Tuvalu, Tonga
	Eric	14–20 Jan	Hurricane	New Caledonia, Vanuatu, Fiji, Tonga
	Nigel	15–28 Jan	Hurricane	New Caledonia, Vanuatu, Fiji
	Odette	17–21 Jan	Hurricane	New Caledonia, Vanuatu
	Freda	26–30 Jan	Hurricane	No land areas
	Gavin	02–09 Mar	Storm	New Caledonia, Vanuatu, Fiji
	Hina	11–20 Mar	Hurricane	New Caledonia, Vanuatu, Fiji
1985–1986	Vernon	23–25 Jan	Gale	Solomon Islands
	Winifred	29 Jan–02 Feb	Hurricane	Australia
	June	05–09 Feb	Storm	French Polynesia
	Ima	05–16 Feb	Hurricane	French Polynesia, Southern Cook Islands
	Keli	08–14 Feb	Gale	New Caledonia, Vanuatu, Fiji, Tonga
	Alfred	03–11 Mar	Gale	Vanuatu, New Caledonia
	Lusi	07–08 Mar		
	Martin	10–15 Apr	Hurricane	Fiji, Tonga, Loyalty Islands
	Manu	23–26 Apr	Storm	Australia
	Namu	16–22 May	Hurricane	Solomon Islands, New Caledonia
1986–1987	Osea	21–25 Nov	Hurricane	No land areas
	Patsy	12–22 Dec	Storm	New Caledonia, Vanuatu
	Raja	21 Dec–05 Jan	Hurricane	Wallis, Futuna, Fiji
	Sally	26 Dec–05 Jan	Hurricane	Cook Islands
	Tusi	15–25 Jan	Hurricane	Tokelau, Samoa
	Uma	04–08 Feb	Hurricane	New Caledonia, Vanuatu
	Veli	05–10 Feb	Gale	Vanuatu, New Caledonia
	Wini	27 Feb–07 Mar	Hurricane	French Polynesia
	Unnamed	28 Feb–02 Mar	Storm	French Polynesia

(continued)

APPENDIX 1. (continued)

Season	Name of cyclone[b]	Duration	Maximum intensity	Places affected[c]
	Yali	07–11 Mar	Hurricane	New Caledonia, Vanuatu
	Zuman	22–26 Apr	Storm	Samoa, Niue
	Blanche	20–26 May	Gale	New Caledonia
1987–1988	Agi	03–14 Jan	Storm	New Caledonia
	Anne	05–14 Jan	Hurricane	New Caledonia, Vanuatu
	Charlie	19 Feb–01 Mar	Storm	New Caledonia
	Bola	24 Feb–11 Mar	Hurricane	New Caledonia, Vanuatu
	Cilla	27 Feb–08 Mar	Hurricane	Vanuatu, French Polynesia
	Dovi	09–18 Apr	Storm	New Caledonia, Vanuatu
1988–1989	Eseta	17–28 Dec	Storm	New Caledonia, Vanuatu, Fiji
	Delilah	31 Dec–08 Jan	Storm	New Caledonia, Vanuatu
	Fili	02–09 Jan	Storm	Samoa, Niue
	Gina	06–09 Jan	Gale	Samoa
	Harry	08–22 Feb	Hurricane	New Caledonia, Vanuatu
	Unnamed	11–14 Feb	Gale	Tonga
	Hinano	21–28 Feb	Hurricane	French Polynesia
	Judy	22–28 Feb	Hurricane	Southern Cook Islands, French Polynesia
	Ivy	23 Feb–03 Mar	Hurricane	New Caledonia, Vanuatu
	Kerry	29 Mar–04 Apr	Storm	Fiji, Tonga
	Aivu	01–04 Apr	Hurricane	Papua New Guinea, Australia
	Lili	07–11 Apr	Hurricane	New Caledonia, Vanuatu
	Meena	03–09 May	Gale	Australia, New Caledonia
	Ernie	09–12 May	Gale	No land areas
1989–1990	Felicity	14–21 Dec	Storm	No land areas
	Nancy	28 Jan–07 Feb	Storm	New Caledonia
	Ofa	30 Jan–10 Feb	Hurricane	Tuvalu, Samoa, Tonga, Niue
	Peni	13–18 Feb	Hurricane	Cook Islands
	Hilda	04–09 Mar	Storm	No land areas
	Rae	16–25 Mar	Storm	Fiji, Tonga
	Ivor	16–25 Mar	Hurricane	No land areas
1990–1991	Sina	24 Nov–04 Dec	Hurricane	Fiji, Tonga, Niue
	Unnamed	14–17 Dec	Gale	Tonga
	Joy	17–26 Dec	Hurricane	Papua New Guinea, Australia
	Kelvin	25 Feb–05 Mar	Storm	Papua New Guinea, Australia
	Lisa	07–19 May	Storm	New Caledonia, Vanuatu
1991–1992	Tia	15–21 Nov	Hurricane	Vanuatu, Solomon Islands
	Val	04–16 Dec	Hurricane	Tokelau, Samoa, Tonga
	Wasa	06–13 Dec	Hurricane	Northern Cook Islands, French Polynesia
	Arthur	14–18 Dec	Storm	French Polynesia
	Betsy	06–19 Jan	Hurricane	Solomon Islands, Vanuatu, New Caledonia
	Cliff	05–11 Feb	Storm	French Polynesia
	Daman	14–23 Feb	Hurricane	Vanuatu, New Caledonia, Australia

(continued)

APPENDIX 1. (continued)

Season	Name of cyclone[b]	Duration	Maximum intensity	Places affected[c]
	Esau	25 Feb–09 Mar	Hurricane	Vanuatu, New Caledonia
	Fran	05–21 Mar	Hurricane	Wallis, Futuna, Fiji, Vanuatu, New Caledonia, Australia
	Gene	15–19 Mar	Storm	Southern Cook Islands
	Hettie	25–29 Mar	Gale	French Polynesia
	Innis	28 Apr–06 May	Storm	Solomon Islands, Vanuatu
1992–1993	Joni	05–14 Dec	Hurricane	Tuvalu, Fiji, Tonga
	Kina	26 Dec–06 Jan	Hurricane	Solomon Islands, Fiji, Tonga
	Nina	29 Dec–05 Jan	Hurricane	Solomon Islands, Tuvalu, Fiji, Wallis, Futuna, Tonga
	Lin	31 Jan–04 Feb	Hurricane	Samoa, Niue
	Mick	05–11 Feb	Gale	Tonga, Fiji
	Oliver	05–14 Feb	Hurricane	Solomon Islands, Papua New Guinea, Australia
	Nisha	12–16 Feb	Storm	Southern Cook Islands, French Polynesia
	Oli	15–20 Feb	Gale	Fiji
	Polly	25 Feb–09 Mar	Hurricane	New Caledonia
	Roger	12–27 Mar	Hurricane	New Caledonia, Australia
	Prema	26 Mar–06 Apr	Hurricane	Vanuatu, New Caledonia
	Adel	14–16 May	Gale	Papua New Guinea
1993–1994	Rewa	28 Dec–23 Jan	Hurricane	Solomon Islands, Papua New Guinea, New Caledonia, Vanuatu, Australia
	Sarah	19 Jan–04 Feb	Hurricane	Vanuatu, New Caledonia
	Theodore	22 Feb–03 Mar	Hurricane	Vanuatu, New Caledonia
	Tomas	20–27 Mar	Hurricane	Solomon Islands, Vanuatu, New Caledonia, Fiji
	Usha	24 Mar–04 Apr	Storm	Solomon Islands, Vanuatu, New Caledonia
1994–1995	Vania	13–18 Nov	Storm	Vanuatu, New Caledonia
	William	31 Dec–05 Jan	Storm	Southern Cook Islands, French Polynesia
	Violet	03–08 Mar	Hurricane	New Caledonia
	Agnes	17–21 Apr	Hurricane	No land areas
1995–1996	Yasi	15–20 Jan	Gale	Tonga
	Celeste	27–31 Jan	Hurricane	New Caledonia, Australia
	Dennis	16–18 Feb	Gale	Australia
	Ethel	07–13 Mar	Storm	No land areas
	Zaka	09–11 Mar	Gale	New Caledonia
	Atu	11–17 Mar	Gale	New Caledonia
	Beti	22 Mar–02 Apr	Hurricane	New Caledonia
1996–1997	Cyrill	22–27 Nov	Gale	Papua New Guinea, New Caledonia
	Fergus	24 Dec–01 Jan	Hurricane	Solomon Islands, Vanuatu, New Caledonia

(continued)

APPENDIX 1. (continued)

Season	Name of cyclone[b]	Duration	Maximum intensity	Places affected[c]
	Drena	03–13 Jan	Hurricane	Vanuatu, New Caledonia
	Evan	13–19 Jan	Hurricane	Samoa
	Freda	24 Jan–05 Feb	Storm	Fiji
	Gillian	10–13 Feb	Gale	Australia
	Harold	16–24 Feb	Storm	New Caledonia
	Ita	23–24 Feb	Gale	Australia
	Noname	23 Feb–02 Mar	Storm	No land areas
	Gavin	03–14 Mar	Hurricane	Tuvalu, Fiji
	Justin	06–29 Mar	Hurricane	Australia, Papua New Guinea
	Hina	12–21 Mar	Hurricane	Futuna, Fiji, Tonga
	Ian	16–19 Apr	Gale	Fiji, Tonga
	June	02–11 May	Storm	Fiji
	Keli	07–17 Jun	Hurricane	Tuvalu, Wallis, Futuna, Tonga, Southern Cook Islands
1997–1998	Lusi	09–12 Oct	Storm	Solomon Islands, Fiji
	Martin	31 Oct–05 Nov	Hurricane	Northern Cook Islands, French Polynesia
	Nute	18–20 Nov	Storm	Solomon Islands, New Caledonia
	Osea	24–28 Nov	Hurricane	Northern Cook Islands
	Pam	06–10 Dec	Storm	Cook Islands
	Ron	02–06 Jan	Hurricane	Samoa, Wallis, Tonga
	Susan	03–10 Jan	Hurricane	Solomon Islands, Vanuatu, Fiji
	Katrina	03–24 Jan	Hurricane	Solomon Islands, Vanuatu
	Tui	26–27 Jan	Gale	Samoa
	Ursula	29 Jan–02 Feb	Storm	French Polynesia
	Veli	01 Feb–03 Feb	Storm	French Polynesia
	Wes	01 Feb–04 Feb	Storm	Solomon Islands
	Yali	19–29 Mar	Hurricane	Vanuatu, New Caledonia
	Nathan	21–26 Mar	Storm	Australia, Papua New Guinea
	Zuman	30 Mar–10 Apr	Hurricane	Vanuatu, New Caledonia
	Alan	21–25 Apr	Gale	Northern Cook Islands, French Polynesia
	Bart	29 Apr–01 May	Gale	French Polynesia
1998–1999	Cora	23–30 Dec	Hurricane	Wallis, Futuna, Fiji, Tonga
	Dani	15–26 Jan	Hurricane	Vanuatu, New Caledonia
	Rona	10–12 Feb	Hurricane	Australia
	Olina	19–24 Jan	Storm	No land areas
	Pete	20–26 Jan	Storm	No land areas
	Ella	11–13 Feb	Gale	Vanuatu, Solomon Islands, New Caledonia
	Frank	18–27 Feb	Hurricane	New Caledonia
	Gita	27 Feb–02 Mar	Gale	No land areas
	Hali	13–18 Mar	Hurricane	Southern Cook Islands
1999–2000	Iris	07–10 Jan	Hurricane	New Caledonia, Vanuatu, Fiji
	Jo	24–31 Jan	Storm	Vanuatu, Fiji
	Kim	24 Feb–02 Mar	Hurricane	French Polynesia

(continued)

APPENDIX 1. (continued)

Season	Name of cyclone[b]	Duration	Maximum intensity	Places affected[c]
	Steve	27 Feb–01 Mar	Storm	Australia
	Leo	06–10 Mar	Storm	No land areas
	Mona	08–13 Mar	Hurricane	Tonga
	Tessie	01–02 Apr	Storm	Australia
	Vaughan	03–06 Apr	Storm	Australia
	Neil	15–19 Apr	Gale	Fiji
2000–2001	Rona	09–12 Feb	Hurricane	Australia
	Oma	20–23 Feb	Storm	Southern Cook Islands
	Paula	26 Feb–08 Mar	Hurricane	Vanuatu, Fiji, Tonga
	Rita	28 Feb–06 Mar	Storm	French Polynesia
	Sose	05–11 Apr	Storm	Vanuatu, New Caledonia
2001–2002	Trina	30 Nov–01 Dec	Gale	Southern Cook Islands
	Vicki	24 Dec	Gale	No land areas
	Waka	29 Dec–01 Jan	Hurricane	Wallis, Futuna, Tonga
	Claudia	11–13 Feb	Hurricane	No land areas
	Des	05–07 Mar	Storm	New Caledonia
	Upia	26–28 May	Gale	Papua New Guinea
2002–2003	Yolande	05 Dec	Gale	No land areas
	Zoe	26 Dec–01 Jan	Hurricane	Solomon Islands, Vanuatu
	Ami	12–15 Jan	Hurricane	Tuvalu, Fiji, Tonga
	Beni	25 Jan–01 Feb	Hurricane	Vanuatu
	Cilla	27–29 Jan	Gale	Tonga
	Dovi	06–11 Feb	Hurricane	Niue, Cook Islands
	Erica	11–16 Mar	Hurricane	Solomon Islands
	Eseta	11–14 Mar	Hurricane	Tonga
	Fili	14–15 Apr	Gale	No land areas
	Gina	05–09 Jun	Hurricane	Solomon Islands, Vanuatu
2003–2004	Heta	02–07 Jan	Hurricane	Tokelau, Wallis and Futuna, Tonga, Samoa, Niue
	Ivy	23–28 Feb	Hurricane	Vanuatu
	Grace	21–24 Mar	Storm	No land areas
2004–2005	Judy	24–27 Dec	Gale	French Polynesia
	Kerry	05–14 Jan	Hurricane	Vanuatu
	Lola	31 Jan–02 Feb	Gale	Tonga
	Meena	03–08 Feb	Hurricane	Cook Islands
	Nancy	12–17 Feb	Hurricane	Cook Islands
	Olaf	13–20 Feb	Hurricane	Samoa, American Samoa
	Percy	24 Feb–05 Mar	Hurricane	Tokelau, Cook Islands
	Rae	05–06 Mar	Gale	No land areas
	Sheila	22 Apr	Gale	No land areas
2005–2006	Tam	12–14 Jan	Gale	Northern Tonga
	Urmil	13–15 Jan	Storm	Northern Tonga

(continued)

APPENDIX 1. (continued)

Season	Name of cyclone[b]	Duration	Maximum intensity	Places affected[c]
	Jim	27 Jan–01 Feb	Hurricane	New Caledonia, Loyalty Islands
	Vaianu	11–16 Feb	Hurricane	Fiji, Tonga
	Wati	19–26 Mar	Hurricane	No land areas

[a]Includes all tropical cyclones that developed in the South Pacific Ocean east of 160°E and also those that formed in the Coral Sea and Solomon Sea between 140°E and 160°E.

[b]Tropical cyclones that originate east of 160°E are named by the Regional Specialized Meteorological Centre – Tropical Cyclone Centre in Nadi, Fiji, whereas those that originate west of 160°E are named by the Tropical Cyclone Warning Centre in Brisbane, Australia. The RSMC-Nadi and the TCWC-Brisbane use different approved lists of cyclone names, so this is why there is a non-alphabetical sequence of cyclone names in most seasons.

[c]Refers to nations, territories or island archipelagoes affected by strong winds. Additional island groups lying farther from the track may have experienced intense rainfall.

APPENDIX 2. List and description of major river floods generated by tropical cyclones in the Fiji Islands between 1840 and 1997. Summarised from original source (Fiji Meteorological Service undated).

Year and month of flood	Tropical cyclone name (or of strength early unnamed storms)	Parts of Fiji worst affected (or path of cyclone)	Flood description and worst effects
1840 Feb	Hurricane	Rewa and northernViti Levu	'High' flood covering the Rewa River flats (Derrick 1946)
1866 Mar	Hurricane	Eye passed between Viti Levu and Vanua Levu	Cotton plantations inundated on the Rewa River flats; food gardens and coffee plantations buried in silt (Derrick 1946)
1869 Mar	Hurricane		Navua River flood left all flat land 'nearly smooth and with 12 inches of deposit in some places' (Derrick 1951)
1871 Mar	Hurricane	Entire Fiji group, centre over western Fiji	Several drownings in the Ba River (Yeo 1998)
1879 Dec	Hurricane	Entire Fiji group, especially northwestern Fiji	Lautoka town flooded (Blong 1994)
1904 Jan	Hurricane	Central Fiji	Rewa badly affected; 1.8 m inundation at Navua (Blong 1994)
1904 Feb	Hurricane		Floods in the Rewa River (D'Aubert 1994)

(continued)

APPENDIX 2. (continued)

Year and month of flood	Tropical cyclone name (or of strength early unnamed storms)	Parts of Fiji worst affected (or path of cyclone)	Flood description and worst effects
1908 Mar	Hurricane	West and south Viti Levu	Two drownings in the Ba River (Yeo 1998)
1910 Mar	Hurricane	Vanua Levu, Viti Levu, Ovalau	Ba River flood (Yeo 1998). Major flooding in the Rewa River (D'Aubert 1994)
1912 Jan	Hurricane	Entire Fiji group	'High' floods in the Labasa and Rewa rivers (Blong 1994)
1929 Dec	Hurricane	Rotuma and eastern Viti Levu	Rewa River flooded all the plains with 7 fatalities and 150 people rescued. Severe flooding along the Navua delta and plains with 10 fatalities. In Labasa floodwater extended 22.5 km inland with 3 fatalities (Derrick 1951)
1931 Feb–Mar	Hurricane	Northern Vanua Levu, western Viti Levu, southern Lomaiviti group; track looped near Yasawa islands	Looping track near Yasawas produced 2 flood peaks on 21 Feb and 1–2 Mar. Worst recorded floods in many rivers. Flood peak on the night of 21 February, particularly in Rewa, Ba, Lautoka and Sigatoka rivers, with considerable loss of life and damage to crops and property (Fiji National Archives 1931). Ba River records highest ever flood with more than 126 drownings (Yeo 1998). Maximum flood in Ba town was 8.5 m a.m.s.l. (Public Works Department 2000). Rewa River rose by approximately 7.6 m (D'Aubert 1994), probably to its highest point in 40 years with widespread inundation. Many cattle lost (Derrick 1951). Sigatoka River – major floods rose 14.3 m at Nawamagi, carrying away houses, people and livestock; bridge destroyed in Sigatoka town. Listed fatalities 206 (Fiji National Archives 1931), although actual number probably higher
1939 Jan	Hurricane	Viti Levu and Kadavu	Flood damage to roads and bridges countrywide (D'Aubert 1994)
1941 Feb	Hurricane	Lau group, Lomaiviti group, Viti Levu	1.8 m inundation at Navua (Blong 1994) and severe flooding in the surrounding area (D'Aubert 1994)

(continued)

APPENDIX 2. (continued)

Year and month of flood	Tropical cyclone name (or of strength early unnamed storms)	Parts of Fiji worst affected (or path of cyclone)	Flood description and worst effects
1941 Dec		Vanua Levu and Taveuni	Navua River flood (Fiji Meteorological Service 1997d)
1943 Jan	Hurricane	Vanua Levu and Lau group	Flood damage recorded countrywide (Fiji Meteorological Service 1997d)
1948 Jan–Feb	Hurricane	Rotuma, western Viti Levu, Kadavu	Viti Levu rivers in high flood (D'Aubert 1994)
1950 Feb–Mar	Storm	Vanua Levu, north-western Viti Levu	Floods in the Labasa area (Blong 1994)
1952 Jan	Hurricane	Northern Yasawas, western Viti Levu, coincided with another minor cyclone in southeastern Vanua Levu and Lau Group	Labasa River flood (Blong 1994). Severe flooding in the Rewa River and its Wainibuka tributary
1954 Jan	Hurricane	Northwestern Fiji	Some flood damage on Viti Levu (Fiji Meteorological Service 1997d)
1956 Jan	Minor	Approached Fiji from the northwest, passed western Fiji	Rewa River – valleys of the Wainimala and Wainibuka tributaries badly affected; estimates that 33% of the food crops were destroyed; 1 m of water in Nausori town (Blong 1994). Ba River – entire Ba town flats flooded up to 3 m (Yeo 1998); severe damage to sugar crops. Nadi River – considerable flooding; up to 1.5 m of water in cane fields; 3 fatalities reported (Fiji National Archives 1956)
1956 Mar	Minor to moderate	Southern Viti Levu, Kadavu	Severe flooding on Viti Levu (Fiji Meteorological Service 1997d)
1958 Nov–Dec	Hurricane	Rotuma, Yasawas, northern and eastern Fiji	Rewa River rose to 0.6–1.0 m in the Nausori area (D'Aubert 1994)
1964 Dec	Hurricane	Western Viti Levu	Roads and villages flooded, communications disrupted, food crops damaged (D'Aubert 1994)
1965 Feb	Hurricane; one of the largest most intense cyclones on record in Fiji	Northern Vanua Levu, western Viti Levu, northern Yasawas, southwestern Kadavu	Very slow movement of the cyclone delivered prolonged and heavy rain. Widespread flooding on the main islands caused 11 fatalities and heavy stock and crop losses (D'Aubert 1994). Ba River flooded Ba town to a maximum 6.5 a.m.s.l. (Public Works Department 2000). Sigatoka River – major flood (Parry 1987)

(continued)

APPENDIX 2. (continued)

Year and month of flood	Tropical cyclone name (or of strength early unnamed storms)	Parts of Fiji worst affected (or path of cyclone)	Flood description and worst effects
1972 Oct	TC Bebe	Entire Fiji	Rewa River – rose 5.5 m above normal at Nausori town (Blong 1994). Ba River rose to a maximum 5.3 m a.m.s.l. (Public Works Department 2000) and flooded Ba town to 1.5 m (Blong 1994). Sigatoka River – disastrous flood in the valley. Nadi River – Nadi town flooded to a depth of 2.4 m, vehicles washed away and about 500 houses damaged. Lautoka city – flooding to 0.6–1.2 m
1978 Dec	TC Fay	Rotuma, eastern Vanua Levu, Taveuni, Lau group	Major flooding in Vanua Levu and Taveuni (Fiji Meteorological Service 1997d)
1979 Mar	TC Meli	Southern Fiji, especially south-eastern Viti Levu	Major flooding around Viti Levu (Fiji Meteorological Service 1997d)
1980 Apr	TC Wally	Yasawas, southern Viti Levu, Vatulele, Kadavu, Beqa	Almost all the coastal rivers on the south and east of Viti Levu experienced severe flooding, causing extensive damage to the main highway. Considerable loss of livestock and crops (Fiji Meteorological Service 1997d). Most hydro-metric stations destroyed or severely damaged. Sigatoka River – major flooding of the valley (Parry 1987)
1982 Jan	TC Hettie	Mamanuca group and western Viti Levu	Extensive flooding on Viti Levu and Vanua Levu (Fiji Meteorological Service 1983). Maximum flood level in Nadi town of 5.9 m a.m.s.l. (Public Works Department 2000). Sigatoka River rose 4 m at Nacocolevu. Flooding in many areas of Labasa (Blong 1994)
1983 Feb–Mar	TC Oscar	Viti Levu, Yasawas, Lomaiviti group, southern Lau group, Kadavu	Nadi River – 3.7 m of water in the Nadi town market (Blong 1994). Sigatoka River flood height between 7 and 9 m at Nacocolevu, some houses were washed away
1984 Mar	TC Cyril	Western Fiji	Significant flooding on Vanua Levu and Viti Levu. Maximum flood level in Nadi town of 5.6 m a.m.s.l

(continued)

APPENDIX 2. (continued)

Year and month of flood	Tropical cyclone name (or of strength early unnamed storms)	Parts of Fiji worst affected (or path of cyclone)	Flood description and worst effects
1985 Jan	TC Eric	Yasawas, Mamanucas, whole of Viti Levu and southern Lau	Flood damages amounted to approximately F$ª64 million for the country (Raucher 2000). Crop and livestock losses in Western Viti Levu (Fiji Meteorological Service 1997d). Maximum flood level in Nadi town of 4.6 m a.m.s.l. (Public Works Department 2000)
1985 Jan	TC Nigel	Southern Yasawas, Mamanucas, northern Viti Levu, Lomaiviti, southern Lau group	Maximum flood level in Nadi town of 4.7 m a.m.s.l. (Public Works Department 2000). Localised flooding in Lautoka (Blong 1994)
1985 Mar	TC Gavin	Western and southwestern Viti Levu	Severe widespread flooding in Viti Levu; closure of several roads bridges and Nausori Airport. Serious flooding in the valleys of the Ba, Nadi, Sigatoka and Rewa rivers (Fiji Meteorological Service 1986, 1997d). Maximum flood levels of 3.5 m and 3.8 m a.m.s.l. in Sigatoka and Nausori towns (Public Works Department 2000)
1985 Mar	TC Hina	Yasawas, Mamanuca group, Viti Levu, Lomaiviti group	Ba River – maximum flood in Ba town 5.0 m a.m.s.l.. Nadi River – maximum flood level in Nadi town of 5.4 m a.m.s.l. (Public Works Department 2000); 1–2 m of flood water in Nadi town (Blong 1994). Flooding in the Sigatoka and Rewa valleys (Fiji Meteorological Service 1986)
1986 Apr	TC Martin	Vanua Levu, Taveuni, northern Lau group, Vanuabalavu	Parts of Ba town flooded (Yeo 1998); Lautoka – surrounding villages flooded (Blong 1994). Nadi River – maximum flood level of 6.5 a.m.s.l. m. in Nadi town. Sigatoka River – maximum flood level of 3.5 m in Sigatoka town a.m.s.l. (Public Works Department 2000)

(continued)

APPENDIX 2. (continued)

Year and month of flood	Tropical cyclone name (or of strength early unnamed storms)	Parts of Fiji worst affected (or path of cyclone)	Flood description and worst effects
1986 Apr	TC Martin	Trough associated with TC Martin drifted back and became stationary over eastern Fiji	Rewa River – widespread flooding and landslides caused eight fatalities and extensive damage in the tributaries (Fiji Meteorological Service 1987); maximum flood level of 5.0 m a.m.s.l. recorded in Nausori town (Public Works Department 2000)
1986–1987 Dec–Jan	TC Raja	Rotuma, northeastern Vanua Levu, Lau group, Koro	Labasa town's main street under a metre of water for the first time in 57 years (Blong 1994)
1988 Feb	TC Bola	Mamanucas, southwest Viti Levu, Kadavu	Flooding in the Labasa area (Blong 1994)
1988 Dec	TC Eseta	Yasawas, Mamanucas, southwestern Viti Levu	Ba River rose to the level of the Ba bridge. Widespread flooding in Fiji's Northern Division (Blong 1994)
1989 Mar–Apr	TC Kerry	Western Viti Levu, Kadavu	Ba River – rose to bridge level on Ba road (Blong 1994)
1990 Mar	TC Rae	Most of Fiji	Rewa, Ba, Nadi and Sigatoka rivers – maximum flood levels of 4.3 m a.m.s.l. in Nausori town, 5.5 m in Ba town, 5.9 m in Nadi town and 3.0 m in Sigatoka town (Public Works Department 2000). Three fatalities and closure of many roads and bridges countrywide (Fiji Meteorological Service 1997d)
1990 Nov	TC Sina	Southern Viti Levu and Lau	Flood damages approximately F$33 million for the country (Raucher 2000)
1992 Dec	TC Joni	Yasawas, Mamanucas, southwestern Viti Levu, Kadavu	Flooding of rivers in Viti Levu especially the Rewa delta; maximum flood level of 3.66 m a.m.s.l. in Nausori town (Public Works Department 2000). Significant loss of livestock (Fiji Meteorological Service 1997d). Damages approximately F$2 million for the country (Raucher 2000)

(continued)

APPENDIX 2. (continued)

Year and month of flood	Tropical cyclone name (or of strength early unnamed storms)	Parts of Fiji worst affected (or path of cyclone)	Flood description and worst effects
1993 Jan	TC Kina	Yasawas, southern Vanua Levu, northern and eastern Viti Levu, Lomaiviti group, southern Lau group	Prolonged heavy rain combined with high tides and heavy seas, blocking mouths of major rivers, resulting in extensive flooding. Rewa River – peak flood rivals world records for a basin of this size; peak discharge at Nausori town estimated at 18,000 m^3 s^{-1}, 6 m a.m.s.l. (highest on record Public Works Department 2000), with 15,000 m^3 s^{-1} flowing under the Rewa bridge and the remainder as sheet flow over the flood plain; the Wainimala tributary overtopped the Vunidawa bridge by some 16 m. Ba River – flood peak just 0.2 m below the 1931 record; entire Ba town on flat flooded to a maximum level of 6.7 m a.m.s.l.; 1 fatality and approximately F\$13.7 million damage in the Ba area (Yeo 1998, 2000). Sigatoka River – maximum flood level of 4.8 m a.m.s.l. in Sigatoka town, the highest on record; destruction of the Ba and Sigatoka bridges. Almost complete loss of crops in the Sigatoka, Navua and Nausori areas, and major loss of livestock (Fiji Meteorological Service 1997d). Overall flood damage for the country approximately F\$188 million (Raucher 2000)
1993 Feb	TC Oli	Yasawas, Mamanucas, southern Viti Levu, Kadavu, Ono-I-Lau	Ba River – Navatu flats flooded (Yeo 1998). Damage to bridges in the Ba and Sigatoka areas (Fiji Meteorological Service 1997d)
1993 Feb	Trough of low pressure linked to TC Polly		Ba River – maximum flood in Ba town was 6.1 m a.m.s.l. Nadi River – maximum flood level of 7.1 m a.m.s.l. in Nadi town, the second highest flood record (Public Works Department 2000). Three fatalities and significant damage to crops property and disruption to transportation

(continued)

APPENDIX 2. (continued)

Year and month of flood	Tropical cyclone name (or of strength early unnamed storms)	Parts of Fiji worst affected (or path of cyclone)	Flood description and worst effects
1994 Nov	TC Vania	Traversed to the west of Fiji	Severe flooding in Tailevu area caused over 100 ha of crops, horses and cattle worth over F$250,000 to perish (Fiji Meteorological Service 1995)
1997 Jan–Feb	TC Evan and TC Freda		Several low bridges and roads under water with an Irish crossing in Nadi washed away
1997 Mar	TC Gavin	Yasawas, Mamanucas, western Viti Levu	Ba River – maximum flood in Ba town 6.3 m a.m.s.l.; entire Ba town on flats flooded; one fatality and approximately F$6 million damage in the Ba area (Yeo 1998, 2000). Nadi River – maximum flood level of 6.7 m a.m.s.l. in Nadi town. Sigatoka River – maximum flood level of 3.4 m a.m.s.l. in Sigatoka town (Public Works Department 2000). Severe flooding in Labasa (Fiji Meteorological Service 1997d). Overall damages for the country amounted to F$35 million (Raucher 2000)
1997 May	TC June	Taveuni, northern Vanua Levu	Taveuni island – localised flooding on the northern coast; local reports of stream levels some of the highest in living memory (Terry and Raj 1999). Cost of damages for the country approximately F$1 million (Raucher 2000)

[a]1F$ (1 Fiji dollar) = US$0.58 and €0.44; January 2007 exchange rates, not adjusted.

Index